CELL BIOLOGY RESEARCH PROGRESS

ADIPOGENESIS

SIGNALING PATHWAYS, MOLECULAR REGULATION AND IMPACT ON HUMAN DISEASE

CELL BIOLOGY RESEARCH PROGRESS

Additional books in this series can be found on Nova's website under the Series tab.

Additional e-books in this series can be found on Nova's website under the e-book tab.

CELL BIOLOGY RESEARCH PROGRESS

ADIPOGENESIS

SIGNALING PATHWAYS, MOLECULAR REGULATION AND IMPACT ON HUMAN DISEASE

YUNFENG LIN
AND
XIAOXIAO CAI
EDITORS

New York

For permission to use material from this book please contact us:
Telephone 631-231-7269; Fax 631-231-8175
Web Site: http://www.novapublishers.com

Library of Congress Cataloging-in-Publication Data

ISBN: 978-1-62808-750-5

Library of Congress Control Number: 2013945715

Published by Nova Science Publishers, Inc. † New York

Contents

Preface **vii**

Chapter I Molecular Determination of Adipogenic versus Osteogenic Differentiation in Mesenchymal Stem Cells **1**
Alan Nguyen, Le Chang, Michelle A. Scott and Aaron W. James

Chapter II LMNA-Linked Lipodystrophies: Experimental Models to Unravel the Molecular Mechanisms **35**
Arantza Infante, Garbiñe Ruiz de Eguino, Andrea Gago and Clara I. Rodríguez

Chapter III Adipogenesis and Osteoblastgenesis toward the Ageing Intervention in the 21st Century **95**
Takeshi Imai

Chapter IV Role of Wnt/β-Catenin Signaling in Bone Marrow Adiposity Following Cancer Chemotherapy **109**
Kristen R. Georgiou and Cory J. Xian

Chapter V Role of Reactive Oxygen Species in Adipocyte Differentiation **127**
D. Lettieri-Barbato, K. Aquilano, G. Tatulli and M. R. Ciriolo

Chapter VI Adipogenesis: Signaling Pathways, Molecular Regulation and Clinical Impact **145**
Kristin A. McPhillips and Quanjun Cui

Index **183**

Preface

Adipogenesis is the process of cell differentiation by which preadipocytes become adipocytes. Adipogenesis has been one of the most intensively studied models of cellular differentiation. The exact mechanisms of adipogenesis remain unclear, since it involves abundant gene, growth factors, cytokines and environmental factors.

This book provides an in-depth overview of current knowledge about the subject of adipogenesis in the human body, including what is known about molecular mechanisms (e.g. cytokine- and growth-factor-related). Transcriptional regulators control the expression of target genes by the interaction with cofactors, coactivators, chromatin remodelling complexes and also with general transcriptional machinery. Current data suggests that adipogenesis is regulated by complex signaling pathway interactions involving multiple transcription factors. Many signaling pathways follow an inverse relationship between osteogenic and adipogenic differentiation. To provide better insight into the basic mechanisms of MSC fate determination, it is crucial to understand the various signaling pathways and cytokine interactions that coordinate this process. Further elucidation of this dichotomy may both improve understanding of human disease and speed the realization of MSC mediated tissue engineering.

There are a number of gaps in current knowledge of how adipogenesis actually occurs, and the authors are hopeful that the publication of this book will help researchers in this field to decide where to focus their future efforts. Contributions are therefore sought from anyone who is undertaking research in this area. The book will also provide an overview for surgeons and clinicians who wish to be kept abreast of developments in this fascinating subject.

Chapter I – Mesenchymal stem cells (MSCs) are multipotent, self-renewing cells capable of differentiating into adipocytes, osteoblasts, chondrocyte, and myocytes among other mesenchymal lineages. A theoretical inverse relationship exists between osteogenic (bone-forming) and adipogenic (fat-forming) differentiation. Current data suggests that this relationship is regulated by complex signaling pathway interactions involving multiple transcription factors. Simplistically, the control of osteogenic versus adipogenic differentiation is under the purview of two main transcription factors: peroxisome proliferator-activated receptor-γ (PPARγ) and Runt-related transcription factor 2 (Runx2), which are principally regarded as master regulators of adipogenesis and osteogenesis, respectively. All signaling pathways that modulate osteogenic versus adipogenic lineage commitment and differentiation have effects on PPARγ, Runx2, or both. This review examines a number of signaling pathways that are known critical regulators of the balance of osteogenic versus adipogenic differentiation in MSCs. Many signaling pathways follow an inverse relationship between osteogenic and adipogenic differentiation, including canonical wingless type MMTV integration site (WNT) and Hedgehog (HH) signaling - both of which are generally pro-osteogenic and anti-adipogenic stimuli. However, there are other signaling pathways with more pleotropic effects. Cytokines such as Bone Morphogenic Proteins (BMPs) can be either pro-osteogenic or pro-adipogenic in a context dependent manner. To provide better insight into the basic mechanisms of MSC fate determination, it is crucial to understand the various signaling pathways and cytokine interactions that coordinate this process. Further elucidation of this dichotomy may both improve understanding of human disease and speed the realization of MSC mediated tissue engineering.

Chapter II – Among the diseases closely related to adipocyte homeostasis are the *LMNA*-linked lipodystrophies, which are included in clinical syndromes called laminopathies. The laminopathies are caused by various mutations in the lamin A gene (*LMNA*), the protein products of which gene are the principal components of the nuclear lamina, located primarily on the inner nuclear membrane. Lipodystrophies are a clinically heterogeneous group of disorders characterized by adipose tissue loss and redistribution, either in localized or generalized regions of the body. In addition, most of these disorders are accompanied by, or predispose patients to, metabolic complications such as lipid profile disturbances (hypertriglyceridemia and low high-density lipoprotein (HDL) cholesterol), glucose intolerance, insulin resistance, hypertension, hepatic steatosis, as well as an increased risk of premature atherosclerosis and coronary disease. The molecular pathophysiology underlying these disorders is not completely understood however, mouse models of these human diseases are playing an important role in unravelling their molecular mechanism. Mouse models have recapitulated many of the typical clinical features of human lipodystrophies, such as insulin resistance, hyperglycemia, conduction-system defects, muscular dystrophy and lipodystrophy. Nevertheless, sometimes the mouse models do not mimic some of the features that characterize the human diseases. Recently a number of human disease models have been established in order to overcome this deficiency. These disease models are based on mesenchymal stem cells (MSCs) or on induced pluripotent stem cells (iPSCs), taking advantage of the capacity of these cells to differentiate to certain cell types which are affected in *LMNA*-linked lipodystrophies, such as adipocytes. Thus, these models allow the study of the molecular pathological mechanisms of a given disease in a patient-specific and cell specific context. Importantly, human disease models based on stem cells provide a valuable tool for discovering molecular targets for drug screening with the aim of developing therapeutic strategies to combat diseases like the *LMNA*-linked lipodystrophies.

Chapter III – Population ageing is a major trend with global implications in the 21st century. Increasing Healthy Life Expectancy (HALE) is one of humanity's greatest achievements. Our recent study showed that a number of teeth of Japanese elderly people was associated with cognitive functions, and fat & bone parameters were decreasing in late elderly age. These data implicates the relation between adipogenesis and osteoblastgenesis to achieve increasing HALE. The adipocytes and osteoblasts are differentiated from the mesenchymal stem cells (MSC), suggesting that similar and different mechanisms are existed in MSC differentiation into adipocytes and osteoblasts. Our final goal is how to expand HALE with regulating both adipogenesis and osteoblastgenesis. The major adipogenesis regulator is PPARγ, which is the member of nuclear receptor superfamily and ligand-dependent transcription factor. PPARγ-ablation in mature adipocyte leads to adipocyte death, lipodystrophy in mice. Runx2 is one of the osteoblastgenesis regulators. Runx2 mutation display cleidocranial dysplasia in human and mice, and Runx2 knock out mice showed embryonic lethality due to bone formation defects. The chemicals which regulate adipogenesis (PPARγ) and osteoblastgenesis (Runx2) will be discussed.

Chapter IV – The bone marrow microenvironment is home to haematopoietic and mesenchymal cell populations that regulate bone turnover through complex interactions. The high proliferative capacity of these cell populations makes them susceptible to damage and injury, which alters the steady-state function of the bone marrow environment. Following cancer chemotherapy, irradiation and long-term glucocorticoid use, a fatty marrow cavity is typically observed, whereby reduced bone and increased fat formation of marrow stromal progenitor cells often results in increased marrow fat, reduced bone mass and increased fracture risk. Although the underlying mechanisms remain to be clearly elucidated, recent investigations have suggested a switch in lineage commitment of bone marrow mesenchymal stem cells down the adipogenic lineage at the expense of osteogenic differentiation, following damage caused by treatment regimens. As the Wnt/β-catenin signaling pathway has been recognized as the key mechanism regulating stromal commitment, its involvement in the osteogenic and adipogenic lineage commitment switch under damaging conditions has been of great interest. This chapter will review the effects of chemotherapy treatment regimens on commitment to the adipogenic and osteogenic lineages of bone marrow stromal progenitor cells. It will also summarize the Wnt/β-catenin signaling pathway and its role in stromal cell lineage commitment and recovery after damage, as well as its potential use as a therapeutic target.

Chapter V – Compelling evidence demonstrates a relationship between reactive oxygen species (ROS) production and adipocytes differentiation; however, no clear proofs about the genuine source(s) of ROS during adipogenesis are available. The synchronized initiation of adipogenesis and mitochondrial biogenesis indicates that mitochondria play a pertinent role in the differentiation and maturation of adipocytes. The early stages of mitochondrial biogenesis and adipocytes differentiation are strongly related to enhanced ROS production. On the basis of this evidence, it is likely that mitochondrial-derived ROS could be mainly involved in the initiation of the redox cascade triggering adipocytes differentiation. Intriguingly, ROS are essential to activate the transcriptional machinery necessary to evoke adipogenesis. Here the authors discuss how adequate levels of ROS maintain cellular homeostasis by creating a suitable redox environment that allows and sustains pre-adipocyte differentiation without causing cellular oxidative damage.

Chapter VI – There has been an overwhelming amount of research in adipogenesis over the past several years and it has been cited as one of the most widely studied areas of cellular biology. The interest likely stems in part from the availability of in vitro models that seem to reliably replicate critical aspects of adipogenesis in vivo. The pathways from mesenchymal stem cell to preadipocyte to adipocyte are well defined. Key mediators in stem cell commitment have been identified in the BMP and Wnt families. Fat genes, notably PPAR-γ, have been identified, and regulators of the expression of the adipocyte phenotype in committed preadipocytes have also been identified all along the pathway, from epigenetic histone gene modulators to modifiers of protein folding in the endoplasmic reticulum. There is an increasing focus on the effect of inflammatory markers like IGF-1 and TNF, as well as environmental toxins and exposures, such as organophosphates, arsenic and obesogen exposures in-utero. Compounds like ginseng, ginkgo, selenate and resveratrol are showing promise as down-regulators of adipogenesis. Obesity and its sequelae--diabetes, hypertension and vascular disease, are global epidemics, and further understanding of the regulation of this pathway has the potential to have a profound clinical impact. In addition to its relevance for obesity and the metabolic syndrome, dysregulation of adipogenesis has been implicated in many other clinical problems, from osteoporosis to aging and cancer. Although research on a cellular level and in animal models has shown promise, it has yet to carry over to the bedside. Continued research is needed to further elucidate the roles and modulators of each pathway in order to develop drugs that either reduce adipogenesis or make adipose tissue more thermogenic.

In: Adipogenesis
Editors: Y. Lin and X. Cai

ISBN: 978-1-62808-750-5

Chapter I

Molecular Determination of Adipogenic versus Osteogenic Differentiation in Mesenchymal Stem Cells

Alan Nguyen, Le Chang, Michelle A. Scott and Aaron W. James
Department of Pathology & Laboratory Medicine,
David Geffen School of Medicine, University of California,
Los Angeles, Los Angeles, US

Abstract

Mesenchymal stem cells (MSCs) are multipotent, self-renewing cells capable of differentiating into adipocytes, osteoblasts, chondrocyte, and myocytes among other mesenchymal lineages. A theoretical inverse relationship exists between osteogenic (bone-forming) and adipogenic (fat-forming) differentiation. Current data suggests that this relationship is regulated by complex signaling pathway interactions involving multiple transcription factors. Simplistically, the control of osteogenic versus adipogenic differentiation is under the purview of two main transcription factors: peroxisome proliferator-activated receptor-γ (PPARγ) and Runt-related transcription factor 2 (Runx2), which are principally regarded as master regulators of adipogenesis and osteogenesis, respectively. All signaling pathways that modulate osteogenic versus adipogenic lineage commitment and differentiation have effects on PPARγ, Runx2, or both. This review examines a number of signaling pathways that are known critical regulators of the balance of osteogenic versus adipogenic differentiation in MSCs. Many signaling pathways follow an inverse relationship between osteogenic and adipogenic differentiation, including canonical wingless type MMTV integration site (WNT) and Hedgehog (HH) signaling - both of which are generally pro-osteogenic and anti-adipogenic stimuli. However, there are other signaling pathways with more pleotropic effects. Cytokines such as Bone Morphogenic Proteins (BMPs) can be either pro-osteogenic or pro-adipogenic in a context dependent manner. To provide better insight into the basic mechanisms of MSC fate determination, it is crucial to understand the various signaling pathways and cytokine interactions that coordinate this process. Further elucidation of this dichotomy may both improve understanding of human disease and speed the realization of MSC mediated tissue engineering.

Keywords: Mesenchymal Stem Cells Adipogenesis Osteogenesis Wnt Hedgehog BMP

Introduction

Mesenchymal stem cells (MSCs) are a multipotent, self-renewing population of stromal cells derived from all vascularized organs, most commonly isolated from bone marrow [1]. These non-hematopoietic cells can differentiate into a variety of mesenchymal tissue, such as bone, cartilage, muscle, ligament, and adipose [2], making them highly attractive as a therapeutic platform. Adipose tissue is a common source of MSCs, termed adipose derived stem cells (ASCs), as it is readily accessible with minimal morbidity [3].

However, the stromal vascular fraction of adipose tissue represents a heterogeneous cell population that is not immediately suitable for bone formation [3]. While relatively scarce, MSCs derived from bone marrow (BMSCs) have a capacity for repeated expansion in culture while retaining their growth potential and multipotency [2]. These BMSCs typically express cell markers, CD29, CD44, CD73, CD105 and CD166, and are negative for hematopoietic markers (including CD14, CD34, and CD45) [2, 4]. However, it is worth noting that with the breadth of sources and protocols for derivation,

MSCs remain poorly defined by their phenotypic, physical, and functional properties [5].

Consequently, wide heterogeneity exists in the MSC literature across species, tissue type of derivation, and even culture strain [6]. Upon commitment and differentiation towards a specific lineage, the gene expression of MSCs shifts until the phenotype is characteristic of the target cell (Figure 1). While MSC lineage commitment and differentiation can be directed by mechanical [7], electrical [8, 9], and magnetic [10] stimuli, this review will specifically focus on biochemical stimuli via cytokine signaling.

MSCs function as precursors to a variety of mesenchymal cell types, including adipocytes (Figure 1) [2]. According to Sinal *et al.,* this process of differentiation into adipocytes, or adipogenesis, can be characterized by two phases: the determination phase and the terminal differentiation phase [11]. During the determination phase, multipotent MSCs commit to the adipocyte lineage exclusively. Morphologically, preadipocytes are indistinguishable from their precursor forms. During the terminal differentiation phase, preadipocytes become adipocytes and acquire new functions, including lipid synthesis and storage as well as adipocyte-specific protein production [12].

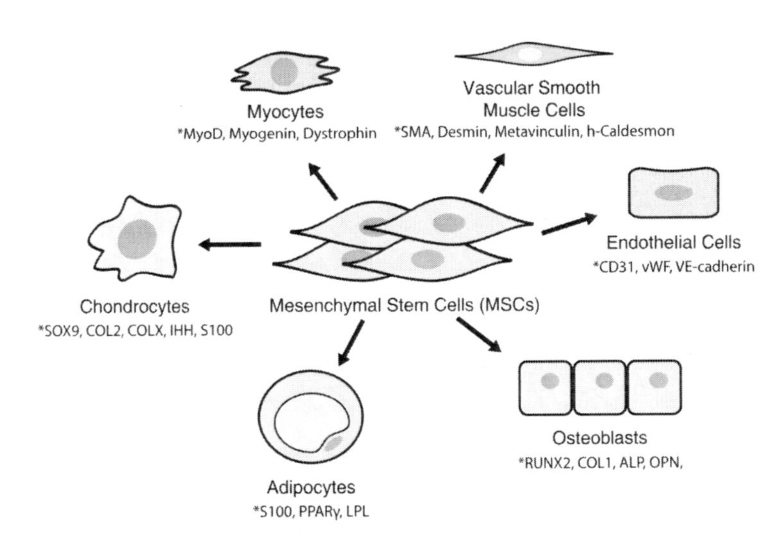

Figure 1. Potential lineages of mesenchymal stem cells (MSCs). MSCs can be derived from all vascularized organs, and their multipotency allows for differentiation into a variety of mesenchymal tissue. These include osteoblasts, adipocytes, chondrocytes, myocytes, vascular smooth muscle cells, and endothelial cells among others. The determination of MSC fate depends on a variety of stimuli, which include mechanical, electrical magnetic, and biochemical factors. *Indicates phenotypic markers for the given lineage.

According to Rosen et al., adipogenesis can be viewed as the shift in gene expression of MSCs that eventually leads to phenotypes that define mature adipocytes [13]. These phenotypes include markers CD24, CD29, CD34, and CD36 [14-16]. The overall process requires a sequentially and temporally ordered, intricate signaling cascade that involves regulatory changes in key transcription factors, most notably Peroxisome proliferator-activated receptor-γ (PPARγ) [11, 13].

Mesenchymal stem cells also give rise to form osteoblasts [2]. The process starts with commitment of osteoprogenitor cells (MSCs) and differentiation into pre-osteoblasts, which eventually develop into mature osteoblasts [17]. Whereas the pathway for adipogenesis requires PPARγ, osteoblast differentiation depends on the key transcription factor, Runt-related transcription factor 2 (Runx2) [17]. However, this alone is not sufficient to become a mature osteoblast, as both osteoblast fate and differentiation are controlled by extracellular signals reviewed in this chapter, including wingless-type MMTV integration site (Wnt), Hedgehog, and Bone Morphogenetic Proteins (BMPs) among others [18]. The phenotypic development of an immature osteoblast into a mature one can be categorized into phases of proliferation, maturation and matrix synthesis, and matrix mineralization [reviewed in [17]]. Osteoblasts synthesize bone matrix to initially form bone and later for bone remodeling, and function in mineral metabolism [18]. Thus far, MSCs have shown potential for osteogenesis *in vitro* [19], and even bone regeneration both in preclinical animal models [reviewed in [20]] and clinical case reports [reviewed in [21, 22]].

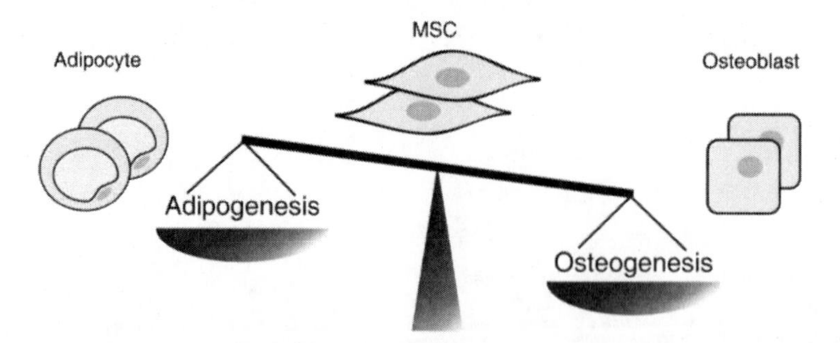

Figure 2. Theoretical Inverse Osteogenic vs. Adipogenic Differentiation in MSCs. Multiple growth factors and other stimuli have been shown to both induce adipogenesis and inhibit osteogenesis, or vice versa. The differentiation of MSCs into either adipocytes or osteoblasts can be theorized as a balance, where preference for one lineage comes at the expense of the other. However, notable exceptions exist to this simplified schematic.

In MSCs, differentiation towards an adipogenic versus an osteogenic lineage depends on a variety of signaling proteins and transcription factors. A large body of experimental evidence suggests an inverse correlation between MSC adipogenesis and osteogenesis (Figure 2) [23, 24].

In vitro, many factors upregulating osteogenesis are associated with a downregulation in adipogenesis, and vice versa [25-28]. In evaluating for these effects, it is important to first recognize that within MSCs there are a few primary cell types.

Although less abundant than other cell types, bone marrow derived stromal cells (BMSCs) are one of the most commonly studied due to its multipotency and retained growth potential after repeated expansion [1]. Adipose tissue derived stem cells (ASCs) are a readily accessible and abundant cell type, with high rate of cell growth and robust potential for osteogenesis both *in vitro* and *in vivo* [3]. Umbilical cord blood-derived mesenchymal stem cells (UC-MSCs) are also readily available, express immunomodulatory molecules, and involve noninvasive means of collection [29, 30].

Several specific cell lines are commonly used for the study of osteo- versus adipogenic differentiation including the pluripotent C3H10T1/2 cell, which display fibroblastic morphology and similar function to MSCs, and murine BMSC line M2-10B4, derived from murine marrow stroma, among others [31, 32]. This inverse relationship is exemplified by the pro-osteogenic and anti-adipogenic wingless type MMTV integration site (Wnt), which does so through both its β-catenin dependent canonical pathway and β-catenin independent non-canonical pathway.

A similar phenomenon is observed with the Sonic Hedgehog (SHH) morphogen, which plays an essential role during skeletogenesis. Most notably, SHH activation promotes osteogenesis while inhibiting adipogenesis in MSCs [23, 33, 34].

However, there are exceptions to this inverse relationship. Bone morphogenetic proteins (BMPs) are well known for their pro-osteogenic effect, and in fact BMP-2 and BMP-7 are commercially available for diverse orthopaedic applications [35, 36].

While the majority of BMPs promote osteogenic commitment and terminal differentiation of MSCs [37, 38], BMPs also have the ability to promote adipogenesis depending on the specific ligand, receptor type, and dose [39, 40]. This chapter will highlight the underlying mechanisms by which Wnt, SHH, and BMPs influence MSC osteo- and adipogenesis.

Clinical Applications of Mesenchymal Cells

The increasing use of MSCs in regenerative medicine and immune intervention has been attributed to their breadth of *in vivo* effects [41]. Although MSCs have a multipotent differentiation capacity, this chapter will cover the clinical uses of adipogenic and osteogenic specific lineages.

MSCs committed to the adipocyte lineage provide cells that contain the important features of mature adipocytes, such as lipolytic capacity and adipokine secretion [42].

Currently, there is a need for mature adipocytes and preadipocytes in a clinical setting as human white adipocyte cell lines are not readily available. The capability of MSCs to differentiate towards adipogenic lineages can be used to treat soft tissue defects after trauma, acute burns and surgery [43]. By engineering adipose tissue, MSCs present a highly promising approach to address the need for innovative therapies to improve subdermal reconstruction.

In addition, clinical trials have already started to use the trophic and immunosuppressive effects of MSCs to help patients suffering from wound healing defects, such as the healing of external fistulas in Crohn's disease [44, 45].

MSCs provide an excellent alternative for engineering bone as bone supply is limited and difficult to collect due to risk of infection, pain during collection and potential loss of function [46]. Surgeons can use scaffolds of synthetic or natural biomaterials, such as hydroxyapatite and tricalcium phosphate, which have osteogenic and osteoinductive properties, to stimulate migration, proliferation and differentiation of stem cells into bone [47].

Currently there is a debate as to whether *in vitro* pre-differentiation is required for MSC mediated bone formation as other sources of MSCs, such as adipose tissue, do not need culture but may have inferior osteogenic potency [48-50]. Currently, clinical studies have used MSCs in a variety of cases to treat patients with osteogenesis imperfecta, reducing complications from orthopaedic surgeries and repairing bone defects [51-53].

Because of the differentiation potential, immunosuppressive properties and cultured transformation of MSCs, there are potential risks involved in their therapeutic use. However, current MSC clinical trials, which number in the hundreds from the US alone (ClinicalTrials.gov), have reported few significant adverse effects.

Transcription Factors

To better understand the effects of signaling proteins, it is important to first establish the key transcription factors found in downstream signaling. One such protein is PPARγ, which exhibits well-understood, pro-adipogenic and anti-osteoblastogenic properties. Conversely, Runx2 is considered the master regulator of osteogenesis. Together, they have been postulated as responsible for mediating the effects of various cytokines and determining the adipogenic versus osteogenic differentiation of MSCs. Further supporting the inverse adipogenic-osteoblastogenic relationship, many studies have found that increased expression of one transcription factor is typically associated with downregulation of the other [54-57]. *Although* PPARγ and Runx2 *are the main transcription factors that mediate adipogenic and osteogenic lineage commitment, there are a variety of other transcription factors such as Osterix and CCAAT/enhancer-binding family of proteins* (C/EBP), which play important adjunctive roles. [See [58, 59] for a comprehensive review of the respective functions of Osterix and (C/EBP) in osteogenesis and adipogenesis]

Peroxisome proliferator-activated receptors are members of the steroid/thyroid hormone receptor gene superfamily [60] and were initially named for PPARα [61]. Subsequently, structural analogs, PPARδ and PPARγ have been discovered and all are activated by polyunsaturated fatty acids [62]. All three PPARs are found in mammals and interact with binding sites on targeted genes by forming heterodimers with the retinoid X receptor (RXR) in order to recruit transcriptional co-activator proteins [63]._Although PPARα and PPARδ are both expressed during adipogenesis, PPARγ is adipocyte cell type restricted and rapidly increases in expression during early adipogenesis [64, 65]. Indeed, PPARγ is considered the master regulator of adipogenesis, for no other factor can rescue adipocyte formation in the event of PPARγ knockout, and generally all cell signaling pathways leading to adipogenesis eventually converge with PPARγ [12, 13, 66]. In fact, $PPAR\gamma^{+/-}$ mice have reduced ability to differentiate into adipocytes, depicting PPARγ as an essential factor in adipogenesis [67, 68]. In addition, $PPAR\gamma^{+/-}$ mice have demonstrated increased bone mass with increased ostoblastogenesis, while having a marked decrease in adipocytes - demonstrating the pro-adipogenic and anti-osteogenic functions of PPARγ [69].

Runx was originally identified as the binding site for polyomavirus enhancer binding protein (PEBP) and was also identified as the Moloney murine leukemia virus enhancer core binding protein [70]. The Runx family has three distinct genes: Runx1, Runx2 and Runx3, which has a varying α subunit with the same β subunit [71, 72]. The DNA binding domain of the Runx family is called *Runt,* which is homologous to the *Runt* sequence in Drosophila. Runx proteins form a heterodimer with transcriptional co-activator core binding factor β (Cbfβ), a co-transcription factor, in order to bind to DNA [71].The Runx family plays a variety of roles in determining stem cell commitment: Runx1 determines hematopoietic stem cell differentiation [73], Runx2 determines osteoblastic and chondrogenic cell differentiation [74], and Runx3 determines epithelial differentiation, neurogenesis and chondrocyte maturation [75, 76]. Runx has also been well studied in its contrasting functions as an oncogene and tumor suppressor. It is currently postulated that Runx family loss of function seems to be a key event in certain myeloid, lymphoid and epithelial cancers [77, 78]. Retroviral insertion of Runx2, which causes transcriptional activation, has also been found to be a potential oncogenetic function of these proteins [79]. However, strong circumstantial evidence suggests that Runx3 is a tumor suppressor as it is methylated and downregulated in cancer derived cell lines and expression of Runx3 inhibits carcinomia cell line growth [80-83]. As the Runx family is structurally similar, it is possible that tissue specific Runx pathways allows for its complex role in cancer. In terms of osteogenesis, Runx2 activates and regulates osteogenic differentiation as a targeted gene of many signaling pathways including but not limited to: Transforming growth factor-beta 1 (TGF-β1), bone morphogenetic protein 2 (BMP-2), Wingless type (Wnt) and Hedgehog [84-86]. Runx2 is necessary for osteogenesis as Runx2 knockout mice lack differentiated osteoblasts and therefore are completely deficient in bone [87]. Runx2 null phenotypes cannot be rescued, however, heterozyote phenotypes can be partially rescued by modulating expression of transcriptional repressor zinc finger protein 521 (Zfp521) and (Nel)–*l*ike protein type *1 (NELL-1) [88, 89]. Although the Runx2 is not directly involved in adipocyte differentiation, its function in promoting osteogenesis may subvert potential adipocyte lineages from occurring in MSCs.*

Wnt Signaling

Since its discovery thirty years ago, wingless-type MMTV integration site (Wnt) signaling has been identified to play a crucial role in cell fate determination, proliferation and differentiation, and tissue homeostasis [90, 91]. In fact, dysregulation of Wnt signaling results in hyperactivation of the pathway and is associated with diseases such as neurodegeneration [92], gastrointestinal cancers [93], and osteoporosis [90], among others. Within the past decade, understanding of the Wnt pathway has rapidly developed with now over 15 receptors and co-receptors identified throughout seven families of proteins [91]. Collectively, Wnt signaling has been demonstrated to have pro-osteogenic, anti-adipogenic properties through utilization of at least two mechanisms: canonical (β-catenin dependent) and non-canonical pathways (Figure 3).

Under the β-catenin dependent pathway, extracellular Wnt ligands bind to frizzled receptors (Frz) at the cell surface. This subsequently induces complex formation with and activation of transmembrane low-density lipoprotein receptor (LRP5/6) coreceptor and intracellular proteins of the disheveled (DSH) family [94]. The resulting complex then inhibits an intracellular complex comprised of Axin, Glycogen synthase kinase 3 (GSK3), and Adenomatosis polyposis coli (APC) protein among others (Fig. III). Of these, GSK3 is responsible for phosphorylation of β-catenin, which itself regulates several transcription factors, thus leading to destabilization. Accordingly, inhibition of the Axin/GSK3/APC complex via canonical Wnt signaling prevents this destabilization, and consequently allows more β-catenin to enter the nucleus (Figure 3). Once inside the nucleus, β-catenin can interact with lymphoid enhancer-binding factor/T cell factor (LEG/TCF) [94]. Ultimately, changes in gene expression via the β-catenin dependent Wnt pathway mediate MSC lineage determination [95] [see [90] for a more comprehensive review of Wnt signaling transduction].

Although the non-canonical Wnt pathway also involves frizzled receptors (Frz) and DSH downstream, it is otherwise independent of β-catenin [96-98]. The non-canonical route can be divided into the planar cell polarity (PCP) pathway and the Ca^{2+} pathway, which collectively involve c-jun NH2-terminal kinase (JNK), calmodulin kinase II (Cam-KII), and protein kinase C (PKC) [98]. Non-canonical agonists, such as Wnt5a, stimulate osteogenesis through activation of this PCP pathway, along with the Ca^{2+} pathway, [99, 100]. Specifically, these effects utilize co-receptor Ror2, which has a demonstrated role in osteogenesis both *in vitro* and *in vivo* [101, 102]. Conversely, Wnt5a inhibition through siRNA has been shown to reduce osteogenic markers in human MSCs [103].

In humans and mice, the relationship between canonical Wnt signaling and bone mass is well established with a key function in promoting osteogenesis. Over the course of the past 8 years, the role of Wnt/β-catenin signaling in bone homeostasis has been solidified, starting with LRP5 mutational studies upstream of β-catenin [104]. Loss of function LRP5 mutations result in pseudo-glioma syndrome with low bone mass phenotype, while gain of function LRP5 mutations exhibit high bone mass (osteoslerotic) phenotype [105-107]. While LRP5 has been proposed to function through β-catenin, a direct role for β-catenin in regulating osteoblast and osteoclast function was verified only recently [108]; these studies indicate that multiple roles exist for β-catenin, in the early stages of osteogenesis and postnatal development [109].

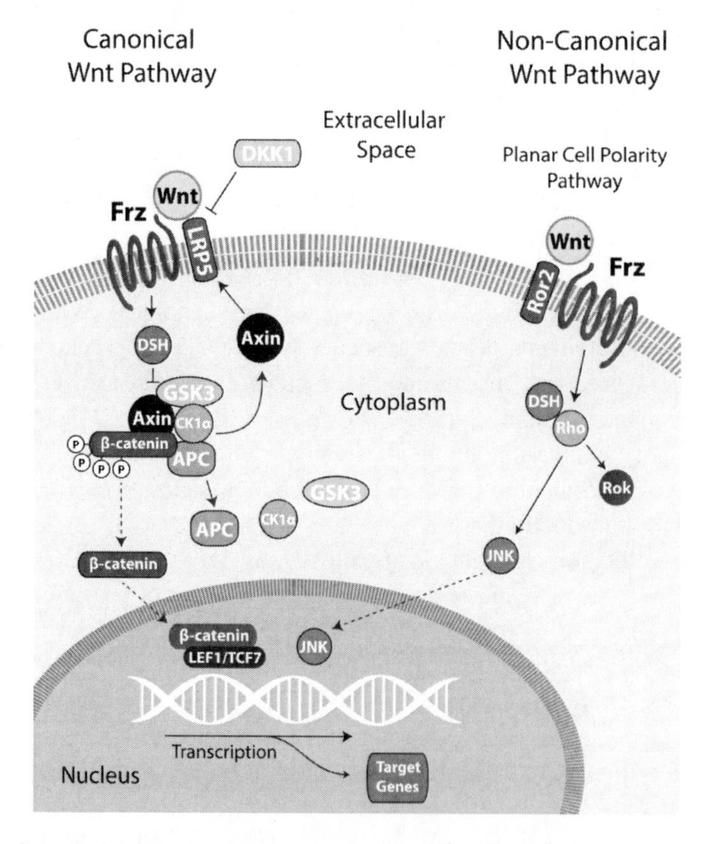

Figure 3. Canonical & Non-canonical Wnt Signaling Pathways. Within the canonical pathway, extracellular Wnt binds to the LRP5-Frizzled (Frz)-disheveled (DSH) complex. This subsequently inhibits the intracellular complex comprised of Axin, Glycogen synthase kinase 3 (GSK3), and Adenomatosis polyposis coli (APC) protein. GSK3 can no longer destabilize β-catenin, and so it is free to enter the nucleus to lymphoid enhancer-binding factor/T cell factor (LEG/TCF) and mediate its osteogenic effects. Under the non-canonical pathway (β-catenin independent), a similar transmembrane complex forms between Wnt, Frz, DSH, and Ror2 for the planar cell polarity (PCP) pathway.

In mesenchymal osteoblastic precursors, for example, β-catenin deficiency resulted in embryonic skeletal defects due to arrest of osteoblast development at an early stage [108, 110-112]. However, in committed osteoblasts, a deficiency in β-catenin resulted in impaired maturation and mineralization, with increased expression of an osteoclast differentiation factor [113, 114]. Thus, disruption of β-catenin in mature osteoblasts has a significant effect on bone resorption.

Recent clinical applications use Wnt inhibitor antagonists to stimulate the Wnt/β-catenin pathway to form new bone and/or inhibit bone resorption. Antagonists sclerostin and Dickkopf-1(Dkk1) downregulate the pathway by binding to and inhibiting Wnt coreceptors LRP5/6 [115]. Inhibitors of these antagonists, anti-sclerostin and anti-Dkk1 respectively, have been shown to stimulate bone formation and increase bone mineral density, with phase I clinical (for anti-sclerostin) and preclinical (for anti-Dkk1) trials currently underway [116, 117].

In regards to regulation of adipogenesis, various members of the Wnt family have been identified to inhibit its early stages [118]. Accordingly, suppression of Wnt signaling via inhibitors, including several members of the dickkopf family (which bind to coreceptor LRP6 to disrupt Wnt-Frz interaction), results in upregulation of adipogenesis [118-120].

Under the canonical Wnt pathway, activation of β-catenin inhibits adipogenesis through suppression of adipogenic transcription factors PPARγ and C/EBPα, both of which induce gene expression that is characteristic of mature adipocytes, in 3T3-L1 preadipocytes [120, 121]. Conversely, induction of PPARγ inhibits β-catenin signaling [120-122].

The suppression of both PPARγ and C/EBPα activity by canonical ligand Wnt3a, among several others, suggests that Wnt/β-catenin (canonical) signaling pathway works to inhibit the activation of both PPARγ and C/EBPα to elicit its anti-adipogenic effects [123]. However, overexpression of PPARγ and/or C/EBPα is not sufficient in rescuing Wnt-inhibited adipogenesis [11, 123].

Wnt signaling follows the inverse relationship between osteo- and adipogenic differentiation of MSCs, as seen in both canonical and non-canonical pathways. With canonical ligands, there is consistent upregulation of osteogenesis associated with decreased adipogenesis [27, 28], while with inhibitors the opposite is seen. For example, lithium chloride, which activates canonical Wnt via inhibition of GSK3b, promotes osteogenesis while suppressing adipogenic differentiation [124, 125].

Similarly, Wnt-10b stimulates osteogenesis *in vivo* to increase bone mass while blocking adipogenesis in preadipocytes *in vitro* via stabilization of free cystolic β-catenin [119, 120, 126]. Canonical ligands Wnt6 and Wnt10a show similar effects in stimulating osteogenesis while also inhibiting adipogenesis and in preadipocytes *in vitro* [127]. Thus, it is expected then that with disruption of β-catenin, which normally enables osteogenic expression, osteoblast maturation is impaired *in vitro* [113, 114] although adipocyte differentiation increases both *in vitro* and *in vivo* [119, 120, 128]. Through a similar downstream mechanism, inhibitors of this β-catenin dependent pathway also support this reciprocal relationship.

For example, DKK-1 secreted by preadipoctyte cells inhibits osteogenesis while promoting adipogenesis and *in vitro* [129], and is correlated with downregulation of cytoplasmic and nuclear β-catenin levels [130]. Aside from the canonical pathway, there is evidence for lineage determination in MSCs via non-canonical Wnt signaling.

In particular, Wnt5a has been shown to suppress pro-adipogenic PPARγ transactivation when co-induced with pro-osteogenic Runx2 in MSCs [11, 131]. Thus, through various ligands and inhibitors, Wnt signaling generally exerts pro-osteogenic and anti-adipogenic effects via either canonical or non-canonical signal transduction.

Hedgehog Signaling

First identified in Drosophila, vertebrates have three homologues of the Drosophila HH protein, Sonic Hedgehog (SHH), Indian Hedgehog (IHH) and Desert Hedgehog (DHH). DHH function is limited to testis organogenesis [132], whereas SHH and IHH are critical during embryogenesis. SHH plays a key role in embryonic development, during skeletogenesis, it is involved in patterning the axial, appendicular and facial skeleton as well as regulating vertebrate organogenesis [133, 134].

Closely related to SHH through gene duplication, IHH regulates chondrogenesis and endochondral bone formation [135]. In fact, disruption of hedgehog (HH) signaling results in severe abnormalities, the most common of which is holoprosencephaly [136].

In regulation of stem cells, SHH is a crucial moderator of cell differentiation and demonstrates pro-osteogenic and anti-adipogenic properties in multiple MSC types [137].

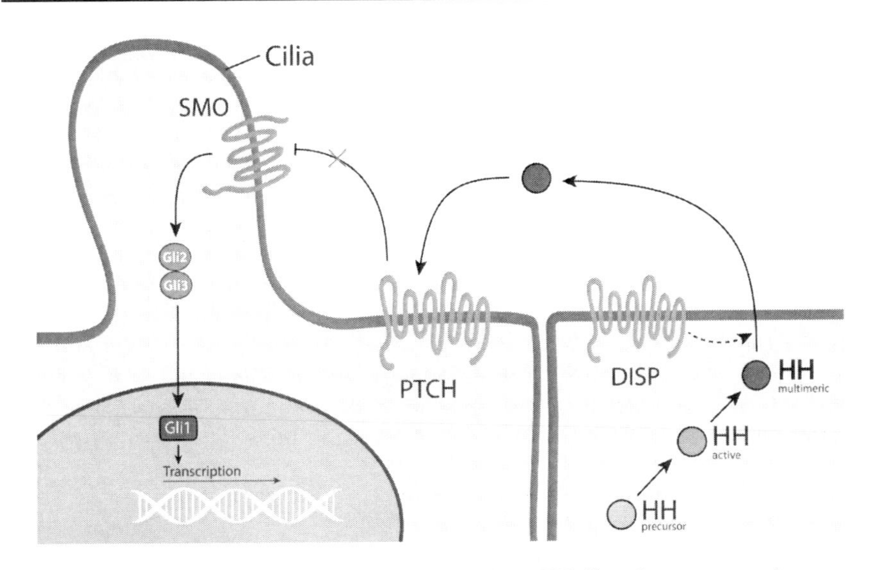

Figure 4. Hedgehog Signaling Pathway. The Hedgehog (HH) ligand precursor undergoes a series of intracellular modifications before reaching an active, multimeric form (shown in yellow). Following release by Dispatched (DISP), the ligand binds to Patched (PTCH), thus releasing Smoothened (SMO) from constitutive inhibition by PTCH. This activates the Gli2/3 complex, which goes on to promote gene expression of Gli1, while repressing the Gli3 repressor that prevents such transcription.

All three HH morphogens utilize the same HH signaling pathway (Fig. IV). First, the HH polypeptide is activated after it is autocatalytically processed from a 45kD to a 19kD protein and is modified at both the C and N-terminus [138]. The modified ligand is secreted by Dispatched on the signaling cell and the HH morphogen attaches the receptor Patched (PTCH), a 12 transmembrane protein, on the receiving cell. After attachment, Smoothened (SMO), a 7 transmembrane protein, is released from suppression, enabling activation of the glioblastoma gene products (Gli), Gli1, Gli2, and Gli3 transcription factors. In fact, Gli1 is a target gene of the HH pathway and has been used as a reliable marker of HH signaling activity [139, 140]. HH signal transduction occurs at the primary cilia and intraflagellar transport proteins are required to produce and preserve cilia during HH signaling [134]. Transfer of PTCH, SMO, and Gli proteins require IFT proteins and Gli activators to move through the cilium to promote genes targeted by HH signaling [141]. Although not fully elucidated, it is believed that HH signaling is mainly mediated though Gli transcription factors and they are responsible for HH-induced lineage commitment during MSC differentiation. Other studies suggest a strong role for oxysterols in utilizing the HH signaling pathway to elicit pro-osteogenic and anti-adipogenic effects in MSCs. Oxysterols elicit osteogenic differentiation in three pathways. First, oxysterol upregulates HH signaling through the promotion of target genes Gli1 and PTCH. Secondly, SMO, downstream from PTCH, is indirectly activated via inhibition of cyclopamine by oxysterols. Finally, the luciferase transporter, found further downstream of SMO, is activated by a Gli-responsive element. Taken together, these findings indicate a role for oxysterols in directing MSCs towards an osteoblastic lineage via activation of the HH pathway [142]. 20(S)-hydroxycholesterol (20S) and its structural analogues are the main oxysterols that induce osteogenesis while suppressing adipogenesis. When oxysterols were used *in vitro*, bMSCs increased osteogenic differentiation and expression of HH target genes: PTCH, Gli1 and Hedgehog Interacting protein (HIP); when used *in vivo*, oxysterols were found to stimulate spinal fusion in murine models [143]. Oxysterols inhibit expression of PPARγ and consequently adipogenesis through an HH-dependent mechanism [144]. However, the suppression of adipocyte formation could be reversed through introduction of cyclopamine, an inhibitor of SMO [145]. It is highly probable that oxysterols, along with the Gli family of transcription factors, enable HH signaling to direct MSC differentiation towards an osteogenic rather than an adipogenic lineage. The anti-adipogenic properties of HH signaling on MSCs have been well documented in a variety of adipocyte and multipotent cell lineages. Adipogenesis in MSCs is characterized by decreased HH signaling as a consequence of decreased Gli1, Gli2, Gli3 and PTCH expression [146]. Conversely, when the HH pathway is activated using a SMO activated inducer of HH signaling such as purmorphamine [147], significant decreases are found in adipocyte-specific markers: including adipocyte fatty acid binding protein (a carrier protein for fatty acids), adipsin, CD36, adiponectin and leptin. By inhibiting expression of adipogenic genes, HH signaling decreases insulin sensitivity, which may reduce the expression of adipogenic transcription factors C/EBPα and PPARγ [146]. Further *in vitro* studies have corroborated the anti-adipogenic function of HH signaling as RNAi scans on Drosophila genome has found that HH blocked differentiation of white adipocytes. In addition, transgenic activation of HH signaling inhibited fat formation in Drosophila and mammalian models [148, 149]. In fact, multipotent C3H10T1/2 cells treated with SHH demonstrated the suppression of the pro-adipogenic effects induced by BMP2 [150]. Another potential method in which SHH inhibits adipogenesis is through the expression of anti-adipogenic transcription factor GATA, which suppresses adipogenesis by binding to promoters of adipogenic factors such as PPARγ [151]. Studies have compared

GATA expression in cells treated with SHH in adipogenic conditions and found that SHH increased expression of GATA2 and GATA3 [152]; however, as these transcription factors have been observed to be down regulated during differentiation, it is unclear if SHH increased GATA expression or the adipocyte differentiation caused the decreasing GATA transcription.

In addition to being anti-adipogenic, HH signaling has been identified to stimulate osteogenic differentiation. Currently, the mechanism and stage at which HH acts during osteoblastogenesis are not well understood. Both *in vivo* and *in vitro* data suggests that HH signaling elicits bone formation via a positive feedback loop: HH-induced osteoblastogenesis require BMP signaling, and together they elicit a synergistic expression of alkaline phosphatase activity [153]. *This positive feedback loop is mediated by Gli2 transcription, which upregulates BMP-2 expression and it is hypothesized that this Gli transcription is also up-regulated by BMP-2 expression [154]. In addition, Gli1 deficient fetuses have exhibited impaired bone formation as well as a lack of both Runx2 and Osterix expression [155].* This indicates that the HH signaling pathway may crosstalk with multiple pathways that regulate osteoblastogenic commitment. For example, SHH demonstrates additive effects alongside the NELL-1 signaling pathway in promoting osteogenesis and repressing adipogenesis [145]. In the murine MSC lineage C3H10T1/2, many studies found that HH simultaneously induced osteoblastic differentiation, while inhibiting adipogenesis [156-158]. In KS483 cells a similar induction of osteogenesis was observed as SHH inhibited adipogenesis despite adipogenic culture conditions [159]. It is important to note that SHH induced differentiation was only observed in immature mesenchymal cell lines 3H10T1/2 c and not pre-osteoblastic MC3T3-E1 or osteoblastic cell lines OS 17/2.8 and ROB-C26 [153, 158]. Therefore, SHH activity may be key in stimulating osteoblastogenesis only during early stages of cell differentiation. In summary, current experimental data corroborates the belief that HH signaling inhibits adipocyte maturation in MSCs while stimulating osteogenesis via the Gli family of transcription factors.

Bone Morphogenetic Protein (BMP) Signaling

Within the transforming growth factor-β (TGF-β) superfamily, Bone Morphogenetic Proteins (BMPs) are extracellular ligands originally isolated as bone extract and identified as having osteogenic properties by their induction of ectopic cartilage and bone formation [160]. They are responsible for numerous processes throughout cell regulation, and ensure proper function and development in tissues such as bone and cartilage [161]. To date, over 20 different BMPs have been identified, of which BMP-2, -4, -6, -7, -9, and -13 are known to contribute to osteogenesis and/or adipogenesis in MSCs [37, 162]. In fact, BMP-2 and -7 were approved by the U.S. FDA in 2002 and 2001, respectively, for regenerative purposes and has transformed the market for spinal fusion surgery and other off-label bone engineering applications [163, 164].

BMPs elicit their effects through interaction with two types of serine-threonine kinases BMP receptors (BMPRs) located on the cell surface. Type II BMP receptors initiate signaling upon binding to a BMP ligand, after which recruitment, phosphorylation and activation of type I BMPRs occurs [161, 165, 166]. While there are several different type I BMPRs, only a few are involved in adipose and skeletal formation, including BMPR-IA and BMPR-IB [39]. After these type I BMPRs are phosphorylated, other signaling cytokines downstream such as Smad1/5/8, MAP Kinase, or c-Jun N-terminal kinase (JNK) signaling pathways are phosphorylated and subsequently activated [39, 167]. Of these, Smads are the most significant cytokine activated by the BMP signaling; it is principally through the Smad-protein complexes that transcription of adipogenic and osteogenic target genes is regulated [161, 165, 166] [see [168] for a more detailed review of BMP signaling transduction].

In regards to adipocyte formation, both Smad and MAP Kinase activation have a significant role [169]. Smad signaling activates PPARγ via zinc finger transcription factor Schnurri-2 and C/EBPα, which exhibit synergistic, adipogenic effects [27, 170]. Accordingly, a Smad antagonist such as Smad6 will interfere with both PPARγ signaling and BMP2-induced adipogenesis [169]. Likewise, disruption of MAP Kinase also suppresses both PPARγ expression and BMP2-induced adipogenesis, whereas activation of MAP Kinase upregulates adipogenesis along with PPARγ activation [169]. BMP signaling further demonstrates an important role as it is identified in the earliest stages of MSC adipogenesis [171, 172]. When MSCs are forced into a preadipocyte cell lineage via exposure to 5-azacytidine, a potent inhibitor of DNA methylation, levels of BMP-4 expression are seen to increase [171, 172]. BMP-4 also shows regular significance in brown adipose tissue, which prioritizes heat production over energy storage (white adipose tissue) [173, 174]. Forced expression of BMP-4 in white adipocytes involving PPARγ coactivator 1-α led to expression of brown adipocyte characteristics, which include increased energy expenditure and insulin sensitivity [175]. Once MSCs have been forced into preadipocyte cells, overexpression of endogenous BMP-4 was sufficient for commitment to the adipocyte lineage [37, 171, 176].

BMP signaling is one of the major pathways involved in the regulation of bone formation. Various murine studies involving genetically modified BMP receptors, BMP ligands, and endogenous inhibitors demonstrate a critical role for BMP in bone formation [177-180]. For example, transgenic mice with modified BMPR-IA receptors exhibit low bone mass and irregular calcification [177]. Inhibitors of BMP signaling, such as Noggin and the Gremlin glycoprotein, impair bone formation when overexpressed [175, 181, 182]. Conversely, Gremlin deletion leads to increased BMP signaling activity (increased Smad phosphorylation) and a subsequent increase in Runx2, ALP activity, and bone formation [180]. In general, BMP regulation of osteogenesis utilizes both autocrine and paracrine pathways [183, 184], and works in conjunction with OSX via both Runx2-dependent and independent pathways. During the process of BMP receptor activation in osteogenesis, as in adipogenesis, both Smads and MAP Kinase induce gene expression downstream. Furthermore, this can be regulated extracellularly by BMP antagonists, such as Noggin or Gremlin, or intracellulary through inhibitory Smads [185, 186].

While over 20 different types of BMPs have been identified, only several actually promote osteogenesis in MSCs. BMP-2, -4, -6, -7, and -9 have been shown to promote osteogenic commitment, as well as terminal osteogenic differentiation of MSCs [37, 38]. BMP-2 treatment leads to osteogenesis for both *in vitro* and *in vivo* studies using MSCs [187-194]. Furthermore, investigators have found that short-term BMP-2 exposure is both necessary and sufficient for irreversible osteogenesis of C3H10T1/2 cells [195]. BMP-9, although less studied is one of the most potent osteogenic BMPs, relying primarily on BMP type I receptors Alk1 and Alk2 receptors [196]. Of the 7 known types of BMP type I receptors, most are present in MSCs [196]. Despite the robust responsiveness of BMP-induced osteogenesis in rat and mice-derived MSCs, results in human MSCs studies have been more variable. For example, several studies evaluating BMP-2, -4, or -7 in human MSCs did not observe increased osteogenesis [197]. However, the variable response of human MSCs to BMPs may reflect overexpression of Noggin by BMP-stimulated hMSCs [198, 199].

The factors that determine if BMP signaling induces adipogenesis or osteogenesis in MSCs are not well understood, but include two known variables: receptor type and dosage. In terms of receptor type, signaling through BMPR-IA in general induces adipogenic effects, while signaling via BMPR-1B induces osteogenic effects. Expression of constitutively active BMPR-IA induces adipogenesis, whereas overexpression of truncated (inactive) BMPR-IA blocks adipogenesis [39]. Conversely, expression of constitutively active BMPR-IB receptors induces mineralization while truncated (inactive) BMPR-IB overexpression blocks osteoblast related genes expression [39]. Consistent with this observation, preadipocyte cells exhibit upregulated levels of BMPR-1A as compared to BMPR-1B [39]. However, there are exceptions to this rule. For example, osteoblast-selective interference of BMPR-IA had unexpected anti-osteogenic effects including irregular calcification and decreased bone mass [177]. Another factor in determining lineage differentiation is BMP dose. In C3H10T1/2 cells, lower concentrations of BMP-2 directs towards adipocyte formation, while higher concentrations favors differentiation into osteoblast [40].

However, high BMP-2 doses in rat femoral defects led to inappropriate adipose tissue formation, although fusion was observed [200]. Thus, while BMP signaling on lineage determination remains under investigation, BMP receptor type and dosage are two known variables that have some effect, although no hard rule applies [109].

Conclusion

There are numerous diverse cytokines involved in the direction of MSC adipogenesis and osteogenesis, some of which will not be covered in this review. Nonetheless, many growth factors influence MSC differentiation in manner that obeys an 'inverse relationship' between adipogenesis and osteogenesis. As discussed, both Wnt and HH signaling follow this pattern, exhibiting pro-osteogenic and anti-adipogenic effects. Other cytokines have similar properties, including Fibroblast Growth Factor-2 (FGF-2) [201], TGF-β1 [84, 202], Nel-like molecule-1 (NELL-1) [203], and Notch signaling [204], to name a few. Other transcription factors have similar pro-osteogenic, anti-adipocytic functions, including the recently described transcriptional activator TAZ (transcriptional activator with PDZ binding motif) [205]. Importantly, there are exceptions to this pattern. For example, both insulin-like growth factor-1 (IGF-1) and BMPs have well documented pro-osteogenic *and* pro-adipocytic properties [195, 206-208]. In summary, a principally inverse relationship exists between adipogenic and osteogenic lineage differentiation in MSCs, observed across multiple MSC cell types. Better elucidation of this dichotomy may both improve understanding of human disease and speed the realization of current efforts in MSC-mediated tissue engineering.

References

[1] Jackson, W.M., L.J. Nesti, and R.S. Tuan, Concise review: clinical translation of wound healing therapies based on mesenchymal stem cells. *Stem Cells Transl. Med.* 1(1): p. 44-50.

[2] Chamberlain, G., et al., Concise review: mesenchymal stem cells: their phenotype, differentiation capacity, immunological features, and potential for homing. *Stem Cells*, 2007. 25(11): p. 2739-49.

[3] James, A.W., et al., Perivascular stem cells: a prospectively purified mesenchymal stem cell population for bone tissue engineering. *Stem Cells Transl. Med.* 1(6): p. 510-9.

[4] Sammons, J., et al., The role of BMP-6, IL-6, and BMP-4 in mesenchymal stem cell-dependent bone development: effects on osteoblastic differentiation induced by parathyroid hormone and vitamin D(3). *Stem Cells Dev*, 2004. 13(3): p. 273-80.

[5] Ho, A.D., W. Wagner, and W. Franke, Heterogeneity of mesenchymal stromal cell preparations. *Cytotherapy*, 2008. 10(4): p. 320-30.

[6] Augello, A. and C. De Bari, The regulation of differentiation in mesenchymal stem cells. *Hum. Gene Ther.* 21(10): p. 1226-38.

[7] Kelly, D.J. and C.R. Jacobs, The role of mechanical signals in regulating chondrogenesis and osteogenesis of mesenchymal stem cells. *Birth Defects Res. C Embryo Today*. 90(1): p. 75-85.

[8] Hronik-Tupaj, M., et al., Osteoblastic differentiation and stress response of human mesenchymal stem cells exposed to alternating current electric fields. *Biomed. Eng. Online*. 10: p. 9.

[9] Creecy, C.M., et al., Mesenchymal stem cell osteodifferentiation in response to alternating electric current. *Tissue Eng. Part A*. 19(3-4): p. 467-74.

[10] Yan, J., et al., Effects of extremely low-frequency magnetic field on growth and differentiation of human mesenchymal stem cells. *Electromagn. Biol. Med.* 29(4): p. 165-76.

[11] Muruganandan, S., A.A. Roman, and C.J. Sinal, Adipocyte differentiation of bone marrow-derived mesenchymal stem cells: cross talk with the osteoblastogenic program. *Cell Mol. Life Sci*, 2009. 66(2): p. 236-53.

[12] Rosen, E.D. and O.A. MacDougald, Adipocyte differentiation from the inside out. *Nat. Rev. Mol. Cell Biol*, 2006. 7(12): p. 885-96.

[13] Rosen, E.D., et al., Transcriptional regulation of adipogenesis. *Genes Dev*, 2000. 14(11): p. 1293-307.

[14] Berry, R. and M.S. Rodeheffer, Characterization of the adipocyte cellular lineage in vivo. *Nat. Cell Biol*. 15(3): p. 302-8.

[15] Jumabay, M., et al., Endothelial differentiation in multipotent cells derived from mouse and human white mature adipocytes. *J. Mol. Cell Cardiol*. 53(6): p. 790-800.

[16] Vroegrijk, I.O., et al., Cd36 is important for adipocyte recruitment and affects lipolysis. Obesity (Silver Spring).

[17] Neve, A., A. Corrado, and F.P. Cantatore, Osteoblast physiology in normal and pathological conditions. *Cell Tissue Res*. 343(2): p. 289-302.

[18] Watanabe, K. and K. Ikeda, [Osteoblast differentiation and bone formation]. Nihon Rinsho, 2009. 67(5): p. 879-86.

[19] Das, A. and E. Botchwey, Evaluation of angiogenesis and osteogenesis. *Tissue Eng. Part B Rev*. 17(6): p. 403-14.

[20] Arinzeh, T.L., Mesenchymal stem cells for bone repair: preclinical studies and potential orthopedic applications. *Foot Ankle Clin*, 2005. 10(4): p. 651-65, viii.

[21] Satija, N.K., et al., Mesenchymal stem cell-based therapy: a new paradigm in regenerative medicine. *J. Cell Mol. Med*, 2009. 13(11-12): p. 4385-402.

[22] Lee, K.D., Applications of mesenchymal stem cells: an updated review. *Chang Gung Med. J*, 2008. 31(3): p. 228-36.

[23] James, A.W., et al., Additive effects of sonic hedgehog and Nell-1 signaling in osteogenic versus adipogenic differentiation of human adipose-derived stromal cells. *Stem Cells Dev*. 21(12): p. 2170-8.

[24] Pei, L. and P. Tontonoz, Fat's loss is bone's gain. *J. Clin. Invest*, 2004. 113(6): p. 805-6.

[25] Beresford, J.N., et al., Evidence for an inverse relationship between the differentiation of adipocytic and osteogenic cells in rat marrow stromal cell cultures. *J. Cell Sci*, 1992. 102 (Pt 2): p. 341-51.

[26] Dorheim, M.A., et al., Osteoblastic gene expression during adipogenesis in hematopoietic supporting murine bone marrow stromal cells. *J. Cell Physiol*, 1993. 154(2): p. 317-28.

[27] Krishnan, V., H.U. Bryant, and O.A. Macdougald, Regulation of bone mass by Wnt signaling. *J. Clin. Invest*, 2006. 116(5): p. 1202-9.

[28] Bennett, C.N., et al., Regulation of osteoblastogenesis and bone mass by Wnt10b. *Proc. Natl. Acad. Sci. U S A*, 2005. 102(9): p. 3324-9.

[29] Robinson, S.N., et al., Mesenchymal stem cells in ex vivo cord blood expansion. *Best Pract. Res. Clin. Haematol*. 24(1): p. 83-92.

[30] Baba, K., et al., Osteogenic potential of human umbilical cord-derived mesenchymal stromal cells cultured with umbilical cord blood-derived fibrin: A preliminary study. *J. Craniomaxillofac. Surg*.

[31] Tang, Q.Q., T.C. Otto, and M.D. Lane, Commitment of C3H10T1/2 pluripotent stem cells to the adipocyte lineage. *Proc. Natl. Acad. Sci. U S A*, 2004. 101(26): p. 9607-11.

[32] Burroughs, J., et al., Diffusible factors from the murine cell line M2-10B4 support human in vitro hematopoiesis. *Exp. Hematol*, 1994. 22(11): p. 1095-101.

[33] Cai, J. and L. Deng, [Regulations of Hedgehog signaling pathway on mesenchymal stem cells]. Zhongguo Xiu Fu Chong Jian Wai Ke Za Zhi. 24(8): p. 993-6.

[34] Cousin, W., et al., Hedgehog and adipogenesis: fat and fiction. *Biochimie*, 2007. 89(12): p. 1447-53.

[35] Jung, R.E., et al., A randomized-controlled clinical trial evaluating clinical and radiological outcomes after 3 and 5 years of dental implants placed in bone regenerated by means of GBR techniques with or without the addition of BMP-2. *Clin. Oral Implants Res*, 2009. 20(7): p. 660-6.

[36] Bessa, P.C., M. Casal, and R.L. Reis, Bone morphogenetic proteins in tissue engineering: the road from laboratory to clinic, part II (BMP delivery). *J. Tissue Eng. Regen Med*, 2008. 2(2-3): p. 81-96.

[37] Kang, Q., et al., A comprehensive analysis of the dual roles of BMPs in regulating adipogenic and osteogenic differentiation of mesenchymal progenitor cells. *Stem Cells Dev*, 2009. 18(4): p. 545-59.

[38] Dorman, L.J., M. Tucci, and H. Benghuzzi, In vitro effects of bmp-2, bmp-7, and bmp-13 on proliferation and differentation of mouse mesenchymal stem cells. *Biomed. Sci. Instrum*. 48: p. 81-7.

[39] Chen, D., et al., Differential roles for bone morphogenetic protein (BMP) receptor type IB and IA in differentiation and specification of

mesenchymal precursor cells to osteoblast and adipocyte lineages. *J. Cell Biol*, 1998. 142(1): p. 295-305.

[40] Wang, E.A., et al., Bone morphogenetic protein-2 causes commitment and differentiation in C3H10T1/2 and 3T3 cells. *Growth Factors*, 1993. 9(1): p. 57-71.

[41] Phinney, D.G. and D.J. Prockop, Concise review: mesenchymal stem/multipotent stromal cells: the state of transdifferentiation and modes of tissue repair--current views. *Stem Cells*, 2007. 25(11): p. 2896-902.

[42] Dicker, A., et al., Functional studies of mesenchymal stem cells derived from adult human adipose tissue. *Exp. Cell Res*, 2005. 308(2): p. 283-90.

[43] Casadei, A., et al., Adipose tissue regeneration: a state of the art. *J. Biomed. Biotechnol*, 2012. 2012: p. 462543.

[44] Ciccocioppo, R., et al., Autologous bone marrow-derived mesenchymal stromal cells in the treatment of fistulising Crohn's disease. *Gut*, 2011. 60(6): p. 788-98.

[45] Garcia-Olmo, D., et al., Treatment of enterocutaneous fistula in Crohn's Disease with adipose-derived stem cells: a comparison of protocols with and without cell expansion. *Int. J. Colorectal Dis*, 2009. 24(1): p. 27-30.

[46] Damien, C.J. and J.R. Parsons, Bone graft and bone graft substitutes: a review of current technology and applications. *J. Appl. Biomater*, 1991. 2(3): p. 187-208.

[47] Niemeyer, P., et al., Evaluation of mineralized collagen and alpha-tricalcium phosphate as scaffolds for tissue engineering of bone using human mesenchymal stem cells. *Cells Tissues Organs*, 2004. 177(2): p. 68-78.

[48] Im, G.I., Y.W. Shin, and K.B. Lee, Do adipose tissue-derived mesenchymal stem cells have the same osteogenic and chondrogenic potential as bone marrow-derived cells? *Osteoarthritis Cartilage*, 2005. 13(10): p. 845-53.

[49] Yoshikawa, T., H. Ohgushi, and S. Tamai, Immediate bone forming capability of prefabricated osteogenic hydroxyapatite. *J. Biomed. Mater Res*, 1996. 32(3): p. 481-92.

[50] Jones, E. and X. Yang, Mesenchymal stem cells and bone regeneration: current status. *Injury*, 2011. 42(6): p. 562-8.

[51] Horwitz, E.M., et al., Isolated allogeneic bone marrow-derived mesenchymal cells engraft and stimulate growth in children with osteogenesis imperfecta: Implications for cell therapy of bone. *Proc. Natl. Acad. Sci. U S A*, 2002. 99(13): p. 8932-7.

[52] Quarto, R., et al., Repair of large bone defects with the use of autologous bone marrow stromal cells. *N. Engl. J. Med*, 2001. 344(5): p. 385-6.

[53] Hernigou, P., et al., Percutaneous autologous bone-marrow grafting for nonunions. Influence of the number and concentration of progenitor cells. *J. Bone Joint Surg. Am*, 2005. 87(7): p. 1430-7.

[54] Valenti, M.T., et al., Role of ox-PAPCs in the differentiation of mesenchymal stem cells (MSCs) and Runx2 and PPARgamma2 expression in MSCs-like of osteoporotic patients. *PLoS One*, 2011. 6(6): p. e20363.

[55] Zhang, L., et al., Melatonin inhibits adipogenesis and enhances osteogenesis of human mesenchymal stem cells by suppressing PPARgamma expression and enhancing Runx2 expression. *J. Pineal Res*, 2010. 49(4): p. 364-72.

[56] Li, X., et al., Lovastatin inhibits adipogenic and stimulates osteogenic differentiation by suppressing PPARgamma2 and increasing Cbfa1/Runx2 expression in bone marrow mesenchymal cell cultures. *Bone*, 2003. 33(4): p. 652-9.

[57] Zhang, X., et al., Runx2 overexpression enhances osteoblastic differentiation and mineralization in adipose--derived stem cells in vitro and in vivo. *Calcif Tissue Int*, 2006. 79(3): p. 169-78.

[58] Zhou, X., et al., Multiple functions of Osterix are required for bone growth and homeostasis in postnatal mice. *Proc. Natl. Acad. Sci. U S A*, 2010. 107(29): p. 12919-24.

[59] Darlington, G.J., S.E. Ross, and O.A. MacDougald, The role of C/EBP genes in adipocyte differentiation. *J. Biol. Chem*, 1998. 273(46): p. 30057-60.

[60] Tontonoz, P. and B.M. Spiegelman, Fat and beyond: the diverse biology of PPARgamma. *Annu. Rev. Biochem*, 2008. 77: p. 289-312.

[61] Issemann, I. and S. Green, Activation of a member of the steroid hormone receptor superfamily by peroxisome proliferators. *Nature*, 1990. 347(6294): p. 645-50.

[62] Gottlicher, M., et al., Fatty acids activate a chimera of the clofibric acid-activated receptor and the glucocorticoid receptor. *Proc. Natl. Acad. Sci. U S A*, 1992. 89(10): p. 4653-7.

[63] Glass, C.K., D.W. Rose, and M.G. Rosenfeld, Nuclear receptor coactivators. *Curr. Opin. Cell Biol*, 1997. 9(2): p. 222-32.

[64] Chawla, A. and M.A. Lazar, Peroxisome proliferator and retinoid signaling pathways co-regulate preadipocyte phenotype and survival. *Proc. Natl. Acad. Sci. U S A*, 1994. 91(5): p. 1786-90.

[65] Adams, M., et al., Activators of peroxisome proliferator-activated receptor gamma have depot-specific effects on human preadipocyte differentiation. *J. Clin. Invest*, 1997. 100(12): p. 3149-53.

[66] Imai, T., et al., Peroxisome proliferator-activated receptor gamma is required in mature white and brown adipocytes for their survival in the mouse. *Proc. Natl. Acad. Sci. U S A*, 2004. 101(13): p. 4543-7.

[67] Kubota, N., et al., PPAR gamma mediates high-fat diet-induced adipocyte hypertrophy and insulin resistance. *Mol. Cell*, 1999. 4(4): p. 597-609.

[68] Barak, Y., et al., PPAR gamma is required for placental, cardiac, and adipose tissue development. *Mol. Cell*, 1999. 4(4): p. 585-95.

[69] Akune, T., et al., PPARgamma insufficiency enhances osteogenesis through osteoblast formation from bone marrow progenitors. *J. Clin. Invest*, 2004. 113(6): p. 846-55.

[70] Wang, S., et al., Cloning and characterization of subunits of the T-cell receptor and murine leukemia virus enhancer core-binding factor. *Mol. Cell Biol*, 1993. 13(6): p. 3324-39.

[71] Bae, S.C. and Y. Ito, Regulation mechanisms for the heterodimeric transcription factor, PEBP2/CBF. *Histol. Histopathol*, 1999. 14(4): p. 1213-21.

[72] Ogawa, E., et al., Molecular cloning and characterization of PEBP2 beta, the heterodimeric partner of a novel Drosophila runt-related DNA binding protein PEBP2 alpha. *Virology*, 1993. 194(1): p. 314-31.

[73] North, T.E., et al., Runx1 is expressed in adult mouse hematopoietic stem cells and differentiating myeloid and lymphoid cells, but not in maturing erythroid cells. *Stem Cells*, 2004. 22(2): p. 158-68.

[74] Yoshida, C.A., et al., Runx2 and Runx3 are essential for chondrocyte maturation, and Runx2 regulates limb growth through induction of Indian hedgehog. *Genes Dev*, 2004. 18(8): p. 952-63.

[75] Levanon, D., et al., The Runx3 transcription factor regulates development and survival of TrkC dorsal root ganglia neurons. *EMBO J*, 2002. 21(13): p. 3454-63.

[76] Brenner, O., et al., Loss of Runx3 function in leukocytes is associated with spontaneously developed colitis and gastric mucosal hyperplasia. *Proc. Natl. Acad. Sci. U S A*, 2004. 101(45): p. 16016-21.

[77] Ito, Y., RUNX genes in development and cancer: regulation of viral gene expression and the discovery of RUNX family genes. *Adv. Cancer Res*, 2008. 99: p. 33-76.

[78] Blyth, K., E.R. Cameron, and J.C. Neil, The RUNX genes: gain or loss of function in cancer. *Nat. Rev. Cancer*, 2005. 5(5): p. 376-87.
[79] Stewart, M., et al., Proviral insertions induce the expression of bone-specific isoforms of PEBP2alphaA (CBFA1): evidence for a new myc collaborating oncogene. *Proc. Natl. Acad. Sci. U S A,* 1997. 94(16): p. 8646-51.
[80] Goel, A., et al., Epigenetic inactivation of RUNX3 in microsatellite unstable sporadic colon cancers. *Int. J. Cancer*, 2004. 112(5): p. 754-9.
[81] Li, Q.L., et al., Transcriptional silencing of the RUNX3 gene by CpG hypermethylation is associated with lung cancer. *Biochem. Biophys. Res. Commun*, 2004. 314(1): p. 223-8.
[82] Kim, T.Y., et al., Methylation of RUNX3 in various types of human cancers and premalignant stages of gastric carcinoma. *Lab. Invest*, 2004. 84(4): p. 479-84.
[83] Guo, W.H., et al., Inhibition of growth of mouse gastric cancer cells by Runx3, a novel tumor suppressor. *Oncogene*, 2002. 21(54): p. 8351-5.
[84] Lee, K.S., et al., Runx2 is a common target of transforming growth factor beta1 and bone morphogenetic protein 2, and cooperation between Runx2 and Smad5 induces osteoblast-specific gene expression in the pluripotent mesenchymal precursor cell line C2C12. *Mol. Cell Biol*, 2000. 20(23): p. 8783-92.
[85] Komori, T., Regulation of osteoblast differentiation by Runx2. *Adv. Exp. Med. Biol*, 2010. 658: p. 43-9.
[86] Pratap, J., et al., Runx2 transcriptional activation of Indian Hedgehog and a downstream bone metastatic pathway in breast cancer cells. *Cancer Res*, 2008. 68(19): p. 7795-802.
[87] Otto, F., et al., Cbfa1, a candidate gene for cleidocranial dysplasia syndrome, is essential for osteoblast differentiation and bone development. *Cell*, 1997. 89(5): p. 765-71.
[88] Hesse, E., et al., Zfp521 controls bone mass by HDAC3-dependent attenuation of Runx2 activity. *J. Cell Biol*, 2010. 191(7): p. 1271-83.
[89] Zhang, X., et al., Nell-1, a key functional mediator of Runx2, partially rescues calvarial defects in Runx2(+/-) mice. *J. Bone Miner. Res*, 2011. 26(4): p. 777-91.
[90] Kim, W., M. Kim, and E.H. Jho, Wnt/beta-catenin signalling: from plasma membrane to nucleus. *Biochem. J*. 450(1): p. 9-21.
[91] Niehrs, C., The complex world of WNT receptor signalling. *Nat. Rev. Mol. Cell Biol*. 13(12): p. 767-79.

[92] Berwick, D.C. and K. Harvey, The importance of Wnt signalling for neurodegeneration in Parkinson's disease. *Biochem. Soc. Trans.* 40(5): p. 1123-8.

[93] White, B.D., A.J. Chien, and D.W. Dawson, Dysregulation of Wnt/beta-catenin signaling in gastrointestinal cancers. *Gastroenterology*. 142(2): p. 219-32.

[94] Pandur, P., D. Maurus, and M. Kuhl, Increasingly complex: new players enter the Wnt signaling network. *Bioessays*, 2002. 24(10): p. 881-4.

[95] Etheridge, S.L., et al., Expression profiling and functional analysis of wnt signaling mechanisms in mesenchymal stem cells. *Stem Cells*, 2004. 22(5): p. 849-60.

[96] Wang, H.Y. and C.C. Malbon, Wnt-frizzled signaling to G-protein-coupled effectors. *Cell Mol. Life Sci*, 2004. 61(1): p. 69-75.

[97] Li, F., Z.Z. Chong, and K. Maiese, Winding through the WNT pathway during cellular development and demise. *Histol. Histopathol*, 2006. 21(1): p. 103-24.

[98] Davis, L.A. and N.I. Zur Nieden, Mesodermal fate decisions of a stem cell: the Wnt switch. *Cell Mol. Life Sci,* 2008. 65(17): p. 2658-74.

[99] Boland, G.M., et al., Wnt 3a promotes proliferation and suppresses osteogenic differentiation of adult human mesenchymal stem cells. *J. Cell Biochem*, 2004. 93(6): p. 1210-30.

[100] Kobayashi, Y., [Roles of Wnt signaling in bone metabolism]. *Clin. Calcium*. 22(11): p. 1701-6.

[101] Liu, Y., et al., Homodimerization of Ror2 tyrosine kinase receptor induces 14-3-3(beta) phosphorylation and promotes osteoblast differentiation and bone formation. *Mol. Endocrinol*, 2007. 21(12): p. 3050-61.

[102] Liu, Y., et al., The orphan receptor tyrosine kinase Ror2 promotes osteoblast differentiation and enhances ex vivo bone formation. *Mol. Endocrinol*, 2007. 21(2): p. 376-87.

[103] Bolzoni, M., et al., Myeloma cells inhibit non-canonical wnt co-receptor ror2 expression in human bone marrow osteoprogenitor cells: effect of wnt5a/ror2 pathway activation on the osteogenic differentiation impairment induced by myeloma cells. *Leukemia*. 27(2): p. 451-63.

[104] Case, N. and J. Rubin, Beta-catenin--a supporting role in the skeleton. *J. Cell Biochem*. 110(3): p. 545-53.

[105] Little, R.D., et al., A mutation in the LDL receptor-related protein 5 gene results in the autosomal dominant high-bone-mass trait. *Am. J. Hum. Genet*, 2002. 70(1): p. 11-9.

[106] Gong, Y., et al., LDL receptor-related protein 5 (LRP5) affects bone accrual and eye development. *Cell*, 2001. 107(4): p. 513-23.
[107] Boyden, L.M., et al., High bone density due to a mutation in LDL-receptor-related protein 5. *N. Engl. J. Med*, 2002. 346(20): p. 1513-21.
[108] Chen, J. and F. Long, beta-catenin promotes bone formation and suppresses bone resorption in postnatal growing mice. *J. Bone Miner. Res.* 28(5): p. 1160-9.
[109] Rahman, S., et al., Œ≤-Catenin Directly Sequesters Adipocytic and Insulin Sensitizing Activities but Not Osteoblastic Activity of PPARŒ≥2 in Marrow Mesenchymal Stem Cells. *PLoS One*. 7(12): p. e51746.
[110] Hu, H., et al., Sequential roles of Hedgehog and Wnt signaling in osteoblast development. *Development*, 2005. 132(1): p. 49-60.
[111] Day, T.F., et al., Wnt/beta-catenin signaling in mesenchymal progenitors controls osteoblast and chondrocyte differentiation during vertebrate skeletogenesis. *Dev. Cell*, 2005. 8(5): p. 739-50.
[112] Hill, T.P., et al., Canonical Wnt/beta-catenin signaling prevents osteoblasts from differentiating into chondrocytes. *Dev. Cell*, 2005. 8(5): p. 727-38.
[113] Holmen, S.L., et al., Essential role of beta-catenin in postnatal bone acquisition. *J. Biol. Chem*, 2005. 280(22): p. 21162-8.
[114] Glass, D.A., 2nd, et al., Canonical Wnt signaling in differentiated osteoblasts controls osteoclast differentiation. *Dev. Cell*, 2005. 8(5): p. 751-64.
[115] Gatti, D., et al., Sclerostin and DKK1 in postmenopausal osteoporosis treated with denosumab. *J. Bone Miner. Res.* 27(11): p. 2259-63.
[116] Lim, V. and B.L. Clarke, New therapeutic targets for osteoporosis: beyond denosumab. *Maturitas*. 73(3): p. 269-72.
[117] Padhi, D., et al., Single-dose, placebo-controlled, randomized study of AMG 785, a sclerostin monoclonal antibody. *J. Bone Miner. Res.* 26(1): p. 19-26.
[118] Laudes, M., Role of WNT signalling in the determination of human mesenchymal stem cells into preadipocytes. *J. Mol. Endocrinol.* 46(2): p. R65-72.
[119] Bennett, C.N., et al., Regulation of Wnt signaling during adipogenesis. *J. Biol. Chem*, 2002. 277(34): p. 30998-1004.
[120] Ross, S.E., et al., Inhibition of adipogenesis by Wnt signaling. *Science*, 2000. 289(5481): p. 950-3.

[121] Liu, J. and S.R. Farmer, Regulating the balance between peroxisome proliferator-activated receptor gamma and beta-catenin signaling during adipogenesis. A glycogen synthase kinase 3beta phosphorylation-defective mutant of beta-catenin inhibits expression of a subset of adipogenic genes. *J. Biol. Chem*, 2004. 279(43): p. 45020-7.

[122] Moldes, M., et al., Peroxisome-proliferator-activated receptor gamma suppresses Wnt/beta-catenin signalling during adipogenesis. *Biochem. J*, 2003. 376(Pt 3): p. 607-13.

[123] Kawai, M., et al., Wnt/Lrp/beta-catenin signaling suppresses adipogenesis by inhibiting mutual activation of PPARgamma and C/EBPalpha. *Biochem. Biophys. Res. Commun*, 2007. 363(2): p. 276-82.

[124] Li, H.X., et al., Roles of Wnt/beta-catenin signaling in adipogenic differentiation potential of adipose-derived mesenchymal stem cells. *Mol. Cell Endocrinol*, 2008. 291(1-2): p. 116-24.

[125] Galli, C., et al., GSK3b-inhibitor lithium chloride enhances activation of Wnt canonical signaling and osteoblast differentiation on hydrophilic titanium surfaces. *Clin. Oral. Implants Res*.

[126] Bennett, C.N., et al., Wnt10b increases postnatal bone formation by enhancing osteoblast differentiation. *J. Bone Miner. Res*, 2007. 22(12): p. 1924-32.

[127] Cawthorn, W.P., et al., Wnt6, Wnt10a and Wnt10b inhibit adipogenesis and stimulate osteoblastogenesis through a beta-catenin-dependent mechanism. *Bone*. 50(2): p. 477-89.

[128] Castro, C.H., et al., Targeted expression of a dominant-negative N-cadherin in vivo delays peak bone mass and increases adipogenesis. *J. Cell Sci*, 2004. 117(Pt 13): p. 2853-64.

[129] Gustafson, B., B. Eliasson, and U. Smith, Thiazolidinediones increase the wingless-type MMTV integration site family (WNT) inhibitor Dickkopf-1 in adipocytes: a link with osteogenesis. *Diabetologia*. 53(3): p. 536-40.

[130] Christodoulides, C., et al., The Wnt antagonist Dickkopf-1 and its receptors are coordinately regulated during early human adipogenesis. *J. Cell Sci,* 2006. 119(Pt 12): p. 2613-20.

[131] Takada, I., et al., A histone lysine methyltransferase activated by non-canonical Wnt signalling suppresses PPAR-gamma transactivation. *Nat. Cell Biol*, 2007. 9(11): p. 1273-85.

[132] Yao, H.H.-C., W. Whoriskey, and B. Capel, Desert Hedgehog/Patched 1 signaling specifies fetal Leydig cell fate in testis organogenesis. *Genes & development*, 2002. 16(11): p. 1433-1440.

[133] Riddle, R.D., et al., Sonic hedgehog mediates the polarizing activity of the ZPA. *Cell*, 1993. 75(7): p. 1401-16.

[134] Ruat, M., et al., Hedgehog trafficking, cilia and brain functions. *Differentiation*, 2012. 83(2): p. S97-104.

[135] Bitgood, M.J. and A.P. McMahon, Hedgehog and Bmp genes are coexpressed at many diverse sites of cell-cell interaction in the mouse embryo. *Dev. Biol*, 1995. 172(1): p. 126-38.

[136] Nanni, L., et al., The mutational spectrum of the sonic hedgehog gene in holoprosencephaly: SHH mutations cause a significant proportion of autosomal dominant holoprosencephaly. *Hum. Mol. Genet*, 1999. 8(13): p. 2479-88.

[137] James, A.W., et al., Sonic Hedgehog influences the balance of osteogenesis and adipogenesis in mouse adipose-derived stromal cells. *Tissue Eng. Part A*, 2010. 16(8): p. 2605-16.

[138] Simpson, F., M.C. Kerr, and C. Wicking, Trafficking, development and hedgehog. *Mech. Dev*, 2009. 126(5-6): p. 279-88.

[139] Hooper, J.E. and M.P. Scott, Communicating with Hedgehogs. *Nat. Rev. Mol. Cell Biol*, 2005. 6(4): p. 306-17.

[140] Ruiz i Altaba, A., C. Mas, and B. Stecca, The Gli code: an information nexus regulating cell fate, stemness and cancer. *Trends Cell Biol*, 2007. 17(9): p. 438-47.

[141] Huangfu, D. and K.V. Anderson, Signaling from Smo to Ci/Gli: conservation and divergence of Hedgehog pathways from Drosophila to vertebrates. *Development*, 2006. 133(1): p. 3-14.

[142] Dwyer, J.R., et al., Oxysterols are novel activators of the hedgehog signaling pathway in pluripotent mesenchymal cells. *J. Biol. Chem*, 2007. 282(12): p. 8959-68.

[143] Johnson, J.S., et al., Novel oxysterols have pro-osteogenic and anti-adipogenic effects in vitro and induce spinal fusion in vivo. *J. Cell Biochem*, 2011. 112(6): p. 1673-84.

[144] Kim, W.K., et al., 20(S)-hydroxycholesterol inhibits PPARgamma expression and adipogenic differentiation of bone marrow stromal cells through a hedgehog-dependent mechanism. *J. Bone Miner. Res*, 2007. 22(11): p. 1711-9.

[145] James, A.W., et al., Additive effects of sonic hedgehog and Nell-1 signaling in osteogenic versus adipogenic differentiation of human adipose-derived stromal cells. *Stem Cells Dev*, 2012. 21(12): p. 2170-8.

[146] Fontaine, C., et al., Hedgehog signaling alters adipocyte maturation of human mesenchymal stem cells. *Stem Cells*, 2008. 26(4): p. 1037-46.

[147] Sinha, S. and J.K. Chen, Purmorphamine activates the Hedgehog pathway by targeting Smoothened. *Nat. Chem. Biol*, 2006. 2(1): p. 29-30.

[148] Pospisilik, J.A., et al., Drosophila genome-wide obesity screen reveals hedgehog as a determinant of brown versus white adipose cell fate. *Cell*, 2010. 140(1): p. 148-60.

[149] Suh, J.M., et al., Hedgehog signaling plays a conserved role in inhibiting fat formation. *Cell Metab*, 2006. 3(1): p. 25-34.

[150] Zehentner, B.K., U. Leser, and H. Burtscher, BMP-2 and sonic hedgehog have contrary effects on adipocyte-like differentiation of C3H10T1/2 cells. *DNA Cell Biol*, 2000. 19(5): p. 275-81.

[151] Tong, Q., J. Tsai, and G.S. Hotamisligil, GATA transcription factors and fat cell formation. *Drug News Perspect*, 2003. 16(9): p. 585-8.

[152] Tong, Q., et al., Function of GATA transcription factors in preadipocyte-adipocyte transition. *Science*, 2000. 290(5489): p. 134-8.

[153] Yuasa, T., et al., Sonic hedgehog is involved in osteoblast differentiation by cooperating with BMP-2. *J. Cell Physiol*, 2002. 193(2): p. 225-32.

[154] Zhao, M., et al., The zinc finger transcription factor Gli2 mediates bone morphogenetic protein 2 expression in osteoblasts in response to hedgehog signaling. *Mol. Cell Biol*, 2006. 26(16): p. 6197-208.

[155] Hojo, H., et al., Gli1 protein participates in Hedgehog-mediated specification of osteoblast lineage during endochondral ossification. *J. Biol. Chem*, 2012. 287(21): p. 17860-9.

[156] Nakamura, T., et al., Induction of osteogenic differentiation by hedgehog proteins. *Biochem. Biophys. Res. Commun*, 1997. 237(2): p. 465-9.

[157] Kinto, N., et al., Fibroblasts expressing Sonic hedgehog induce osteoblast differentiation and ectopic bone formation. *FEBS Lett*, 1997. 404(2-3): p. 319-23.

[158] Spinella-Jaegle, S., et al., Sonic hedgehog increases the commitment of pluripotent mesenchymal cells into the osteoblastic lineage and abolishes adipocytic differentiation. *J. Cell Sci*, 2001. 114(Pt 11): p. 2085-94.

[159] van der Horst, G., et al., Hedgehog stimulates only osteoblastic differentiation of undifferentiated KS483 cells. *Bone*, 2003. 33(6): p. 899-910.

[160] Wozney, J.M., et al., Novel regulators of bone formation: molecular clones and activities. *Science*, 1988. 242(4885): p. 1528-34.

[161] Chen, D., M. Zhao, and G.R. Mundy, Bone morphogenetic proteins. *Growth Factors*, 2004. 22(4): p. 233-41.

[162] Bragdon, B., et al., Bone morphogenetic proteins: a critical review. *Cell Signal*. 23(4): p. 609-20.

[163] Freire, M.O., et al., Antibody-mediated osseous regeneration: a novel strategy for bioengineering bone by immobilized anti-bone morphogenetic protein-2 antibodies. *Tissue Eng. Part A*. 17(23-24): p. 2911-8.

[164] Friedlaender, G.E., et al., Osteogenic protein-1 (bone morphogenetic protein-7) in the treatment of tibial nonunions. *J. Bone Joint Surg. Am*, 2001. 83-A Suppl 1(Pt 2): p. S151-8.

[165] Miyazono, K., S. Maeda, and T. Imamura, BMP receptor signaling: transcriptional targets, regulation of signals, and signaling cross-talk. *Cytokine Growth Factor Rev*, 2005. 16(3): p. 251-63.

[166] Nohe, A., et al., Signal transduction of bone morphogenetic protein receptors. *Cell Signal*, 2004. 16(3): p. 291-9.

[167] Nishimura, R., et al., Regulation of bone and cartilage development by network between BMP signalling and transcription factors. *J. Biochem*. 151(3): p. 247-54.

[168] Li, X. and X. Cao, BMP signaling and skeletogenesis. *Ann. N Y Acad. Sci*, 2006. 1068: p. 26-40.

[169] Hata, K., et al., Differential roles of Smad1 and p38 kinase in regulation of peroxisome proliferator-activating receptor gamma during bone morphogenetic protein 2-induced adipogenesis. *Mol. Biol. Cell*, 2003. 14(2): p. 545-55.

[170] Jin, W., et al., Schnurri-2 controls BMP-dependent adipogenesis via interaction with Smad proteins. *Dev. Cell*, 2006. 10(4): p. 461-71.

[171] Bowers, R.R. and M.D. Lane, A role for bone morphogenetic protein-4 in adipocyte development. *Cell Cycle*, 2007. 6(4): p. 385-9.

[172] Bowers, R.R., et al., Stable stem cell commitment to the adipocyte lineage by inhibition of DNA methylation: role of the BMP-4 gene. *Proc. Natl. Acad. Sci. U S A*, 2006. 103(35): p. 13022-7.

[173] Herzig, S. and C. Wolfrum, Brown and white fat: From signaling to disease. *Biochim. Biophys. Acta*. 1831(5): p. 895.

[174] Hao, R., et al., Brown adipose tissue: distribution and influencing factors on FDG PET/CT scan. *J. Pediatr. Endocrinol. Metab*. 25(3-4): p. 233-7.

[175] Qian, S.W., et al., BMP4-mediated brown fat-like changes in white adipose tissue alter glucose and energy homeostasis. *Proc. Natl. Acad. Sci. U S A*. 110(9): p. E798-807.

[176] Ahrens, M., et al., Expression of human bone morphogenetic proteins-2 or -4 in murine mesenchymal progenitor C3H10T1/2 cells induces

differentiation into distinct mesenchymal cell lineages. *DNA Cell Biol*, 1993. 12(10): p. 871-80.

[177] Mishina, Y., et al., Bone morphogenetic protein type IA receptor signaling regulates postnatal osteoblast function and bone remodeling. *J. Biol. Chem*, 2004. 279(26): p. 27560-6.

[178] Okamoto, M., et al., Bone morphogenetic proteins in bone stimulate osteoclasts and osteoblasts during bone development. *J. Bone Miner. Res*, 2006. 21(7): p. 1022-33.

[179] Gazzerro, E., et al., Conditional deletion of gremlin causes a transient increase in bone formation and bone mass. *J. Biol. Chem*, 2007. 282(43): p. 31549-57.

[180] Gazzerro, E., et al., Skeletal overexpression of gremlin impairs bone formation and causes osteopenia. *Endocrinology*, 2005. 146(2): p. 655-65.

[181] Davis, S.W. and S.A. Camper, Noggin regulates Bmp4 activity during pituitary induction. *Dev. Biol*, 2007. 305(1): p. 145-60.

[182] Zhu, W., et al., Noggin regulation of bone morphogenetic protein (BMP) 2/7 heterodimer activity in vitro. *Bone*, 2006. 39(1): p. 61-71.

[183] Cheng, H., et al., Osteogenic activity of the fourteen types of human bone morphogenetic proteins (BMPs). *J. Bone Joint Surg. Am*, 2003. 85-A(8): p. 1544-52.

[184] Suzawa, M., et al., Extracellular matrix-associated bone morphogenetic proteins are essential for differentiation of murine osteoblastic cells in vitro. *Endocrinology*, 1999. 140(5): p. 2125-33.

[185] Skarzynska, J., et al., Modification of Smad1 linker modulates BMP-mediated osteogenesis of adult human MSC. *Connect Tissue Res*. 52(5): p. 408-14.

[186] Benisch, P., et al., The transcriptional profile of mesenchymal stem cell populations in primary osteoporosis is distinct and shows overexpression of osteogenic inhibitors. *PLoS One*. 7(9): p. e45142.

[187] Reid, J., H.M. Gilmour, and S. Holt, Primary non-specific ulcer of the small intestine. *J. R. Coll. Surg. Edinb*, 1982. 27(4): p. 228-32.

[188] Varkey, M., et al., In vitro osteogenic response of rat bone marrow cells to bFGF and BMP-2 treatments. *Clin. Orthop. Relat. Res*, 2006. 443: p. 113-23.

[189] Partridge, K., et al., Adenoviral BMP-2 gene transfer in mesenchymal stem cells: in vitro and in vivo bone formation on biodegradable polymer scaffolds. *Biochem. Biophys. Res. Commun*, 2002. 292(1): p. 144-52.

[190] Wegman, F., et al., Osteogenic differentiation as a result of BMP-2 plasmid DNA based gene therapy in vitro and in vivo. *Eur. Cell Mater.* 21: p. 230-42; discussion 242.

[191] Park, K.H., et al., Bone morphogenic protein-2 (BMP-2) loaded nanoparticles mixed with human mesenchymal stem cell in fibrin hydrogel for bone tissue engineering. *J. Biosci. Bioeng*, 2009. 108(6): p. 530-7.

[192] Tang, Y., et al., Combination of bone tissue engineering and BMP-2 gene transfection promotes bone healing in osteoporotic rats. *Cell Biol. Int*, 2008. 32(9): p. 1150-7.

[193] Kempen, D.H., et al., Retention of in vitro and in vivo BMP-2 bioactivities in sustained delivery vehicles for bone tissue engineering. *Biomaterials*, 2008. 29(22): p. 3245-52.

[194] Cheng, S.L., et al., In vitro and in vivo induction of bone formation using a recombinant adenoviral vector carrying the human BMP-2 gene. *Calcif Tissue Int*, 2001. 68(2): p. 87-94.

[195] Noel, D., et al., Short-term BMP-2 expression is sufficient for in vivo osteochondral differentiation of mesenchymal stem cells. *Stem Cells*, 2004. 22(1): p. 74-85.

[196] Luo, J., et al., TGFbeta/BMP type I receptors ALK1 and ALK2 are essential for BMP9-induced osteogenic signaling in mesenchymal stem cells. *J. Biol. Chem.* 285(38): p. 29588-98.

[197] Osyczka, A.M., et al., Different effects of BMP-2 on marrow stromal cells from human and rat bone. *Cells Tissues Organs*, 2004. 176(1-3): p. 109-19.

[198] Diefenderfer, D.L., et al., BMP responsiveness in human mesenchymal stem cells. *Connect Tissue Res*, 2003. 44 Suppl 1: p. 305-11.

[199] Chen, C., et al., Noggin suppression decreases BMP-2-induced osteogenesis of human bone marrow-derived mesenchymal stem cells in vitro. *J. Cell Biochem.* 113(12): p. 3672-80.

[200] Zara, J.N., et al., High doses of bone morphogenetic protein 2 induce structurally abnormal bone and inflammation in vivo. *Tissue Eng. Part A.* 17(9-10): p. 1389-99.

[201] Xiao, L., et al., Disruption of the Fgf2 gene activates the adipogenic and suppresses the osteogenic program in mesenchymal marrow stromal stem cells. *Bone.* 47(2): p. 360-70.

[202] Choy, L., J. Skillington, and R. Derynck, Roles of autocrine TGF-beta receptor and Smad signaling in adipocyte differentiation. *J. Cell Biol*, 2000. 149(3): p. 667-82.

[203] James, A.W., et al., A new function of Nell-1 protein in repressing adipogenic differentiation. *Biochem. Biophys. Res. Commun.* 411(1): p. 126-31.

[204] Ugarte, F., et al., Notch signaling enhances osteogenic differentiation while inhibiting adipogenesis in primary human bone marrow stromal cells. *Exp. Hematol*, 2009. 37(7): p. 867-875 e1.

[205] Byun, M.R., et al., TAZ is required for the osteogenic and anti-adipogenic activities of kaempferol. *Bone*. 50(1): p. 364-72.

[206] Wang, S., et al., Insulin-like growth factor 1 can promote the osteogenic differentiation and osteogenesis of stem cells from apical papilla. *Stem. Cell Res*. 8(3): p. 346-56.

[207] Teruel, T., et al., Insulin-like growth factor I and insulin induce adipogenic-related gene expression in fetal brown adipocyte primary cultures. *Biochem. J*, 1996. 319 (Pt 2): p. 627-32.

[208] Huang, Y.L., et al., Effects of insulin-like growth factor-1 on the properties of mesenchymal stem cells in vitro. *J. Zhejiang Univ. Sci. B*. 13(1): p. 20-8.

In: Adipogenesis
Editors: Y. Lin and X. Cai

ISBN: 978-1-62808-750-5

Chapter II

LMNA-Linked Lipodystrophies: Experimental Models to Unravel the Molecular Mechanisms

Arantza Infante, Garbiñe Ruiz de Eguino, Andrea Gago and Clara I. Rodríguez
Stem Cells and Cell Therapy Laboratory, BioCruces Health Research Institute, Hospital Universitario Cruces, Plaza de Cruces S/N, Barakaldo, Bizkaia, Spain

Abstract

Among the diseases closely related to adipocyte homeostasis are the *LMNA*-linked lipodystrophies, which are included in clinical syndromes called laminopathies. The laminopathies are caused by various mutations in the lamin A gene (*LMNA*), the protein products of which gene are the principal components of the nuclear lamina, located primarily on the inner nuclear membrane.

Lipodystrophies are a clinically heterogeneous group of disorders characterized by adipose tissue loss and redistribution, either in localized or generalized regions of the body. In addition, most of these disorders are accompanied by, or predispose patients to, metabolic complications such as lipid profile disturbances (hypertriglyceridemia and low high-density lipoprotein (HDL) cholesterol), glucose intolerance, insulin resistance, hypertension, hepatic steatosis, as well as an increased risk of premature atherosclerosis and coronary disease.

The molecular pathophysiology underlying these disorders is not completely understood however, mouse models of these human diseases are playing an important role in unravelling their molecular mechanism. Mouse models have recapitulated many of the typical clinical features of human lipodystrophies, such as insulin resistance, hyperglycemia, conduction-system defects, muscular dystrophy and lipodystrophy. Nevertheless, sometimes the mouse models do not mimic some of the features that characterize the human diseases. Recently a number of human disease models have been established in order to overcome this deficiency. These disease models are based on mesenchymal stem cells (MSCs) or on induced pluripotent stem cells (iPSCs), taking advantage of the capacity of these cells to differentiate to certain cell types which are affected in *LMNA*-linked lipodystrophies, such as adipocytes. Thus, these models allow the study of the molecular pathological mechanisms of a given disease in a patient-specific and cell specific context. Importantly, human disease models based on stem cells provide a valuable tool for discovering molecular targets for drug screening with the aim of developing therapeutic strategies to combat diseases like the *LMNA*-linked lipodystrophies.

Lipodystrophies

Lipodystrophies defined as a disturbance in adipose tissue, encompass a wide spectrum of disorders, ranging from total wasting of adipose tissue (lipoatrophy), partial loss of adipose tissue (lipodystrophy), to the abnormal accumulation of body fat (lipohypertrophy) (Araujo-Vilar et al., 2012; Garg, 2011; Bidault et al., 2011; Agarwal and Garg, 2006; Bremer et al., 2011; Capeau et al., 2010; Capell and Collins, 2006; Caron et al., 2010). The term lipodystrophy, which means deregulation of adipose tissue, thus includes both lipoatrophy and lipohypertrophy. Furthermore, lipodystrophy should also be considered a major metabolic disorder, rather than simply adipose tissue gain or loss. The dysfunction in adipose tissue is not only limited to its architecture or fat redistribution, but also involves functional anomaly, which will be described below.

Adipose tissue is a major endocrine organ, complex and highly metabolically active (Ahima and Flier, 2000; Fruhbeck et al., 2001), and its alteration can lead to a wide range of diseases. Over the past decade, remarkable breakthroughs have been made in recognizing the complex nature of the adipocyte as a secretory cell as well as a site of the regulation of energy storage regulation (Frayn et al., 2003).

Thus, it is essential to have a broad view of adipose tissue functions. In addition to adipocytes, adipose tissue contains connective tissue matrix, nerve tissue, immune cells (Kershaw and Flier, 2004) and the stromal vascular cell fraction, which is the source of mesenchymal stem cells (MSC) (Mansilla et al., 2011; Riordan et al., 2009; Zuk et al., 2002). Adipokines are a variety of bioactive peptides which are expressed and secreted by the adipose tissue, and which act at both, local (autocrine/paracrine) and systemic (endocrine) levels (Kershaw and Flier, 2004). Examples of adipocyte-derived proteins with endocrine functions include lipid and metabolism related proteins (LPL, CETP, APOE, NEFAs), steroid metabolism enzymes (Cytocrome P450-dep aromatase, 17bHSD, 11bHSD1), cytokines and cytokine-related proteins (leptin, TNFa, IL6), fibrinolitic system proteins (PAI-1, Tissue factor), complement and related proteins (Adiponectin, ASP, complement factor B, adipsin), and proteins of the RAS pathway (AGT) (Kershaw and Flier, 2004). Examples of receptors expressed in adipose tissue include receptors for traditional endocrine hormones (Insuline receptor, Glucagon receptor, GH receptor, TSH receptor, Glucagon like peptide-1 receptor, Angiotensin II receptors type 1 and 2); nuclear hormone receptors (glucocorticoid receptors, vitamin D receptor, Thyroid hormone receptor, Androgen receptor, estrogen receptor, progesterone receptor); cytokine receptors (leptin receptor, IL6 receptor, TNFa receptor), and catecholamine receptors (b1,b2, b3 receptors; a1 y a2 receptors) (Kershaw and Flier, 2004).

In addition, adipose tissue possesses the ability to modulate its own metabolic activities, including differentiation of new adipocytes and angiogenesis if necessary to stockpile increasing fat stores, while signaling other tissues to regulate energy metabolism in accordance with the body's nutritional state (Frayn et al., 2003).

Adipose tissue operates as a structured whole, and its functions are regulated by multiple external influences such as autonomic nervous system activity, blood flow and the delivery of a complex mix of substrates and hormones in the plasma. Through this integrative nature of adipose tissue function and the interactive network, adipose tissue is responsible for coordinating a variety of biological processes including dynamic energy metabolism, neuroendocrine function, and immune function. (Kershaw and Flier, 2004; Frayn et al., 2003). All studies to date indicate that lipodystrophies, due to the fundamental role of adipose tissue in metabolism, are not limited to changes in body fat stores, but are also associated with metabolic abnormalities to varying degrees (Garg, 2011).

The severity of associated complications varies depending on the primary disorder, and the most frequent complications are insulin resistance, lipid profile disturbances (hypertriglyceridemia and low high-density lipoprotein (HDL) cholesterol), glucose intolerance, hypertension, hepatic steatosis as well as an increased risk of premature atherosclerosis and coronary disease (Monajemi et al., 2007; Garg, 2011).

Based on the etiology of the disease, lipodystrophies can be differentiated into two types: the congenital forms are less frequent, while acquired lipodystrophies related to HIV infection are more prevalent. The UNAIDS (Joint United Nations Programme on HIV/AIDS) report on the global AIDS epidemic in 2012 indicated that 8 million people are on antiretroviral therapy, and a total of 34 million people are infected with HIV all over the world.

The estimated prevalence of lypodistrophies ranges from 1 in 200.000 to 1 in 1.500.000, a frequency that is increasing. On one hand HIV infection has gone from being a fatal disease in the short term to a chronic illness because of the treatments available. Although otherwise relatively healthy, infected patients may develop metabolic complications with severe implications on their health (Caron-Debarle et al., 2010). On the other hand, congenital lipodystrophies may offer some treatments that will be able to extend the life of some of these patients (Garg, 2011).

Currently there are symptomatic treatments but no cure for lipodystrophies. There is significant heterogeneity in the pattern of adipose tissue loss among various types of lipodystrophies (Garg, 2011; Mory et al., 2012). Genetically determined and acquired lipodystrophies can be classified into two major categories: partial or generalized disorders, (Agarwal et al., 2006; Jeninga and Kalkhoven, 2010) depending on the extent of body fat loss. If adipose tissue is lost from discrete areas as in the case of autoinflammatory syndromes or repeated subcutaneous injection of drugs, e.g. insulin, it can be classified as the localized variety. In partial lipodystrophy, adipose tissue is lost mainly from the limbs. In generalized lipodystrophic disorders, fat tissue is lost from nearly the entire body (Garg, 2011). A brief classification and description of clinical features of lipodystrophies is provided in Table 1.

Of the many causes of lipodistrophies, genetic defects in one particular gene account for a significant proportion. To date, there are at least five types of lipodystrophies related to *LMNA* gene defects and/or associated proteins. Mutations in *LMNA* are extremely pleiotrophic, causing different phenotypes depending on the mutation in the gene. Even the same *LMNA* mutation can be associated with different forms of the same disease (Garg, 2011; Bidault et al., 2011).

Table 1. Classification of lipodystrophie

Type	Subtype	Gene/Pattern	Main Phenotype	Associated clinical features
Congenital Generalized Lipodystrophy (CGL)	CGL1 (OMIM 608594)	*AGPAT2* / AR	Generalized lipodystrophy at birth	Preservation of mechanical adipose tissue, severe insulin resistance, DMt2, dyslipidemia, acromegaloid features, hepatomegaly, hypoleptinemia
	CGL2 (OMIM 269700)	*BSCL2* / AR	Generalized lipodystrophy at birth	Lack of mechanical adipose tissue, severe insulin resistance, DMt2, dyslipidemia, acromegaloid features, hepatomegaly, mild mental retardation, cardiomyopathy, hypoleptinemia
	CGL3 (OMIM 612526)	*CAV1* / AR	Generalized lipodystrophy.	Insulin resistance, DMt2, dyslipidemia
	CGL4 (OMIM 613327)	*PTRF* / AR	Generalized lipodystrophy.	Muscular dystrophy and in some patients: hepatomegaly, insulin resistance and dyslipidemia
Mandibuloacral Dysplasia (MAD)	Type B (OMIM 608612)	*ZMPSTE24* / AR	Generalized lipodystrophy mandibular hypoplasia and resorption of clavicles	Insulin resistance, DMt2, dyslipidemia, premature aging, growth retardation, chronic renal failure
Hutchinson-Gilford progeria syndrome (HGPS)	(OMIM 176670)	*LMNA* / AD	Generalized lipodystrophy, normal at birth, premature aging	Abnormalities in skin, hair loss, bone and cartilage disorders, insulin resistance, dyslipidemia, premature atherosclerosis and cardiovascular disease
Pubertal Onset Generalized Lipodystrophy	(OMIM 608056)	*LMNA* / AD	Generalized lipodystrophy onset during puberty	Atypical progeroid features: beaked nose, lack of hair, atrophic skin over hands and feet, DMt2 and dyslipidemia
Neonatal Progeroid Syndrome	(OMIM 264090)	NK / probably AR	Generalized lipodystrophy and progeroid appearance at birth	Sparing of gluteus fat, multiple skeletal and eye abnormalities

Table 1. (Continued)

Type	Subtype	Gene/Pattern	Main Phenotype	Associated clinical features
Familial Partial Lipodystrophy (FPL)	FPL1 or Kobberling variety (OMIM 608600)	NK	Partial lipodystrophy, loss of adipose tissue confined to the extremities, with normal or increased distribution of fat on the face, neck, and trunk	Severe hyperlipidemia, eruptive xanthomas, insulin resistance DM with lack of ketoacidosis, hepatomegaly, and increased basal metabolic rate.
	FPLD2 or Dunningan variety (OMIM 151660)	*LMNA* / AD	Partial lipodystrophy onset during puberty	Fat accumulation in face and neck, insulin resistance, DMt2, dyslipidaemia, hepatic steatosis, PCOS, Hypoleptinemia, hypoadiponectinemia, cardiomyopathy and muscular dystrophy in some patients
	FPL3 (OMIM 604367)	*PPARG* / AD	Partial lipodystrophy	Insulin resistance, DMt2, dyslipidaemia, hypertension, PCOS
	FPL4 (OMIM 613877)	*PLIN1* / AD	Partial lipodystrophy, onset during late childhood	Severe dyslipidemia, insulin resistance, DMt2, hypertension, loss of subcutaneous adipose tissue mainly in lower limbs
	FPL5	*AKT2* / AD	Partial lipodystrophy	Insulin resistance, Dt2M, hypertension
	FPL6	*CIDEC* / AR	Partial lipodystrophy	Insulin resistance, DMt2, dyslipidaemia.
Mandibuloacral Dysplasia (MAD)	Type A (OMIM 248370)	*LMNA* / AR	Partial lipodystrophy, mandibular hypoplasia, resorption of clavicles and acro-osteolysis	Mild insulin resistance, DMt2, dyslipidaemia, premature aging, postnatal growth retardation, mottled cutaneous pigmentation, hyperinsulinemia with insulin resistance, DMt2, dyslipidaemia
HIV-Related Lipodystrophy		Several genes / Acquired	Partial lipodystrophy	Insulin resistance, DMt2, dyslipidaemia.

Type	Subtype	Gene/Pattern	Main Phenotype	Associated clinical features
SHORT syndrome	(OMIM 269880)	NK / AD and AR	Partial lipodystrophy, as part of multiple birth anomalies	Short stature, Hyperextensibility and hernias, Ocular depression, Rieger anomaly, Teething delay.
Acquired Partial Lipodystrophy (APL) or Barraquer-Simons Syndrome	(OMIM 608709)	*LMNB2* or NK / Acquired	Partial lipodystrophy, onset before 15 years old	Gradual onset of bilaterally symmetrical loss of subcutaneous fat from the face, neck, upper extremities, trunk, in 'cephalocaudal' sequence, sparing of the lower extremities. Lipohypertrophy in legs and hips. Some patients have low serum complement component C3 and the autoantibody C3 nephritic factor, with or without GNMP. Others have hyperglycaemia, dyslipidemia and hypertension. PCOS.
Werner Syndrome (WNR)	(OMIM 277700)	*RECQL2* / AR	Partial lipodystrophy as part of another syndrome	Scleroderma-like skin changes, especially in the extremities, cataracts, subcutaneous calcification, premature arteriosclerosis, diabetes mellitus, and prematurely aged features.
Atypical Progeroid Syndrome (APS)		*LMNA* / AD or sporadic (*de novo* mutations)	Partial or generalized lipodystrophy with progeroid syndrome	DMt2, mandybular hypoplasia, skin pigmentation.

AR: Autosomal recesive; AD: Autosomal Dominant; NK: not known; DMt2: diabetes mellitus type 2; PCOS : polycystic ovarian syndrome; GNMP: membranoproliferative glomerulonephritis.

The *LMNA* Gene and the Nuclear Lamina

The nuclear lamina is an intermediate filament network composed of proteins called lamins underlying the nuclear envelope of all somatic cells. In vertebrates, two classes of lamins exist: A and B-type lamins, which polymerize to form the nuclear lamina at the nucleoplasmic side of the inner nuclear membrane. A-type lamins (lamin A and C) are both encoded by the *LMNA* gene (Lin and Worman, 1993); whereas, B-type lamins (lamin B1 and

B2) are encoded by the *LMNB1* and *LMNB2* genes respectively (Biamonti et al., 1992; Lin and Worman, 1995). While B type lamins are expressed in all somatic cells, lamins A and C are absent from embryonic stem cells (Stewart and Burke, 1987; Constantinescu et al., 2006).

Although initially the nuclear lamina was thought to mainly provide mechanical stability to the nucleus, many studies in the last decade have implicated lamins in a broad variety of functions. Thus, in addition to its structural role, much evidence indicates that the nuclear lamina modulates gene expression either by directly interacting with chromatin (Verstraeten et al., 2007; Reddy et al., 2008; Lee et al., 2009), or by sequestering transcription factors at the nuclear periphery (Ivorra et al., 2006; Gonzalez et al., 2008). In addition, A-type lamins are scaffolds for proteins that regulate DNA synthesis, DNA damage response, chromatin organization, gene transcription, cell cycle progression, cell migration and cell differentiation (Broers et al., 2006; Verstraeten et al., 2007).

In mammalian somatic cells, A-type lamins are represented by lamins A and C which originate from alternative splicing of the *LMNA* gene and only differ in the carboxyl-terminal domain of the proteins. Prelamin A is the translation product of the mature *LMNA* mRNA in healthy individuals. It terminates with a CaaX motif (where "C" is a cysteine, "a" is often an aliphatic amino acid, and "X" is one of many different residues). The CaaX motif triggers three sequential enzymatic modifications (Zhang and Casey, 1996) (Figure 1). First, a 15-carbon farnesyl lipid is added to the cysteine residue by protein farnesyltransferase (FTase).

After protein farnesylation, which is one form of protein prenylation, the last three amino acids (the –aaX) are clipped off by an endoprotease specific for prenylated proteins. For lamin B1 and many other CaaX proteins, this proteolytic processing step is carried out by an RCE1 homolog, prenyl protein peptidase (RCE1) (Maske et al., 2003). For prelamin A, this step is likely a redundant function of RCE1 and a zinc metallopeptidase STE24 homolog (ZMPSTE24) (Bergo et al., 2002). Next, the newly exposed farnesylcysteine is methylated by a membrane methyltransferase, isoprenylcysteine carboxyl methyltransferase (ICMT) (Dai Q et al., 1998). The last 15 amino acids of prelamin A, including the farnesylcysteine α-methyl ester, are excised by ZMPSTE24 and degraded, leaving mature lamin A (Weber et al., 1989; Beck et al., 1990).

Importantly, this final endoproteolytic processing step does not occur in the absence of ZMPSTE24, indicating that the function of this metallopeptidase is required in the final processing of prelamin A (Bergo et al., 2002; Pendás et al., 2002).

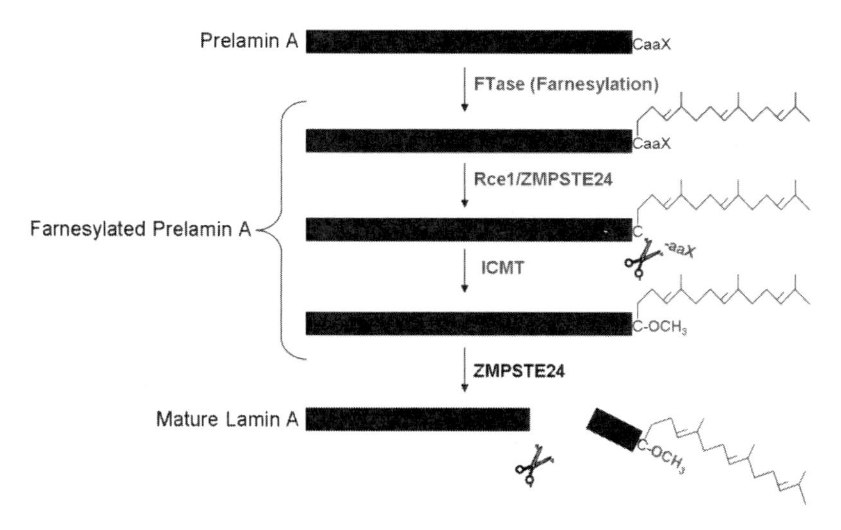

Figure 1. Schematic representation of the posttranslational processing of Prelamin A.

It is thought that this complex posttranslational process is carried out to facilitate nuclear envelope targeting of farnesylated prelamin A, which is subsequently cleaved by ZMPSTE24 to remove the farnesyl groups and produce mature lamin A, which can then insert into the nuclear lamina (Sinensky et al., 1994).

Mutations in the *LMNA* gene or a defective posttranslational processing of prelamin A (mutations in *ZMPSTE24*, f. e.) are responsible for more than ten different diseases called laminopathies, which include systemic disorders and tissue restricted diseases (Capell and Collins, 2006; Verstraeten 2007). In all cases, there is abnormal blebbing of the nuclear membrane, which is regarded a hallmark of these diseases (Capell and Collins, 2006).

Among them, the premature aging syndrome known as Hutchinson-Gilford Progeria syndrome (HGPS) and some partial lipodystrophies share a defective physiological maturation of prelamin A as the main pathophysiological mechanism. Progeria is caused by *de novo* point mutations that interfere with conversion of farnesyl-prelamin A to mature lamin A (De Sandre-Giovannoli et al., 2003).

The *LMNA*-linked lipodystrophies currently include five main groups, Familial partial lipodystrophies (FPLD), Mandibuloacral Dysplasia type A and B (MADA and MADB), Hutchinson–Gilford Progeria Syndrome (HGPS), and

HIV-related lipodystrophy (Jeninga and Kalkhoven, 2010; Caron et al. 2010; Villarroya et al., 2010).

Clinical Features and Main Characteristics of Lamina Related Lypodystrophies

Familial Partial Lipodystrophy, Dunningan Type (FPLD2)

FPLD2, an autosomal dominant disorder, is the most frequent variety of FPL more than 300 patients have been reported (Garg, 2011) since 2000, the year in which Cao and Hegele reported that the disease was caused by a mutation in the *LMNA* gene (Cao and Hegele, 2000).

FPLD2 is characterized by normal subcutaneous adipose tissue distribution until the onset of puberty, when affected individuals gradually start to lose subcutaneous fat from limbs, gluteal and truncal regions (Kobberling and Dunnigan,1986; Garg, 2011). Concurrently adipose tissue accumulates in the face, neck and intraabdominal areas; the patients also present metabolic abnormalities such insulin-resistant, type 2 diabetes mellitus, dyslipidaemia, hepatic steatosis, acanthosis nigricans, hypoleptinemia and hypoadiponectinemia. Moreover, some patients may have cardiomyopathy and muscular dystrophy. (Kobberling and Dunnigan, 1986; Garg, 2011; Agarwal and Garg, 2006; Bidault et al., 2011; Araujo-Vilar et al., 2012). Women are more severely affected, often have Polycystc Ovary Syndrome (PCOS) with hyperandrogenism, and a high risk of infertility, gestational diabetes, and obstetrical complications (Vantyghem et al., 2008).

It is important to note that the pro-atherogenic profile (accumulation of atherogenic risk factors) developed by these patients may lead to cardiovascular diseases related to premature atherosclerosis.

Consistent with the findings of Caron and coworkers (Caron et al., 2007a) that there is an altered processing of prelamin A in adipose tissue from patients affected with lipodystrophy, Araujo-Vilar and collaborators (Araujo-Vilar et al., 2009) demonstrated prelamin A accumulates in the nuclear envelope of peripheral adipose tissue of patients with FPLD2. Ultrastructural analysis showed defects in the heterochromatin peripheral nuclear layer and a nuclear fibrous dense lamina.

In addition to adipocytes, these alterations were present in fibroblasts and endothelial cells of the patients (Caron et al., 2007b). Cultured skin fibroblasts from patients also exhibit increased oxidative stress which triggers premature cellular senescence. In these manners, the accumulation of prelamin A is associated with premature aging complications (Caron et al., 2007a).

Although common to other laminophaties (Bidault et al., 2011), FPLD2-linked mutations induce changes in lamin interactions with chromatin-associated proteins or with DNA, thus modifying chromatin organization (Capanni et al., 2005; Vigoroux et al., 2011). Accumulated prelamin A in lipodistrophic cells have been shown to alter adipogenesis through the sequestration of SREBP1, a transcription factor implicated in adipogenesis and lipid homeostasis, and related with the down-regulation of PPARγ expression (Capanni et al., 2005; Bidault et al., 2011). The impairment of adipogenesis also seems to be mediated by dysfunctional action of pRb (retinoblastoma protein), C/EBPβ (CCAAT enhancer binding protein beta) and LPL (Lipoprotein Lipase) among others (Araujo-Vilar et al., 2009; Araujo-Vilar et al., 2012). However, the mechanisms by which accumulation of prelamin A could interfere with the balance of adipose tissues are complex and may involve other proteins and transcription factors as described recently by our group (Ruiz de Eguino et al., 2012) where prelamin A accumulation impairs hMSC differentiation to adipocytes through its interaction with SP1 transcription factor.

Several heterozygous *LMNA* gene missense mutations have been described which are associated with FPLD2: R482Q (Cao and Hegele, 2000), R482W and R482L (Shackleton et al., 2000), G465D (Speckman et al., 2000), D230N and R399C (Lanktree 2007).

FPLD2 is not the only FPL recognized. Four FPL disorders have been identified with genetic mutations in other genes (Table 1), and there are indications that FLP patients exist with phenotypes of unknown causes (Agarwal and Garg, 2006; Garg, 2011).

Mandibuloacral Dysplasia Type A (MADA)

Mandibuloacral dysplasia type A (MADA) is classified as a partial variety of lipodystrophy, which is a rare autosomal recessive disorder characterized by craniofacial anomalies with mandibular hypoplasia, growth retardation, high-pitched voice, mottled skin pigmentary changes and skeletal abnormalities such as progressive osteolysis of clavicles and distal phalanges, and joint contractures (Agarwal and Garg, 2006; Garg, 2011). Furthermore the loss of acral adipose tissue with normal or excess adipose tissue in the face, neck and truncal regions is a distinctive characteristic of this disease. Some patients may have progeroid features such as skin atrophy, superficial vasculature, alopecia, short stature and thin beaked nose.

Metabolic complications can arise due to insulin resistance, type 2 diabetes mellitus and dyslipidemia (Agarwal and Garg, 2006; Garg, 2011).

Many studies on the molecular basis of MADA emphasize the phenotypic variability in patients with *LMNA* mutations. Novelli and coworkers (Novelli et al., 2002) identified a homozygous missense mutation (R527H) in the *LMNA* gene in 2002.

Further *LMNA* mutations in the C-terminal region of the gene have also been identified, both heterozygous (R472C/R527C) and homozygous (K542N) (Plasilova et al., 2004). The precise mechanism by which defective *LMNA* and the resulting accumulation of prelamin A causes these phenotypic alterations remains unclear.

Among some of the numerous cellular alterations associated with *LMNA* mutations in this disease are chromatin interactions, epigenetic changes, nuclear blebbing, lamina assembly and decreased resistance to mechanical stress (Filesi et al. 2005; Capanni et al., 2012).

Mandibuloacral Dysplasia Type B (MADB)

Mandibuloacral dysplasia with type B lipodystrophy is a generalized form of lipodystrophy, caused by heterozygous mutations in the *ZMPSTE24* gene (Agarwal et al., 2003). *ZMPSTE24* encodes the zinc metalloproteinase involved in the post-translational proteolytic processing of lamin A, which activity alteration results in the accumulation of farnesylated prelamin A in the cells (Agarwal et al., 2003; Garg, 2011). Patients exhibit characteristics similar to the MADA described above.

Distinctive features include premature birth, early onset of manifestations, renal disease, calcified skin nodules and a lack of acanthosis nigricans. This MADB phenotype is also associated with insulin resistance and its attendant metabolic complications (Garg, 2011).

Hutchinson-Gilford Progeria Syndrome (HGPS)

While other lipodystrophies have been recognized for more than a century, the progress from molecular description and unraveling of the molecular causes of Hutchinson-Gilford Progeria Syndrome (HGPS) has presented an amazing evolutionary process. A mere 10 years elapsed between the discovery of a *de novo* dominant C-to-T point mutation in *LMNA* (Erikson et al., 2003) and the completion of the first clinical trial with a possible treatment (Gordon et al., 2012).

The onset of HGPS is usually within the first year of life (Goldman et al., 2004) (Bidault et al., 2011). The children appear normal at birth, but in the first year present a failure to grow, generalized lipodystrophy, very low body weight, characteristic faces with receding mandibles, narrow nasal bridges and pointed nasal tips (resembling aged persons), short stature, early loss of hair, scleroderma, joint contractures, bone changes, nail dystrophy and delayed primary tooth eruption. Cognitive development is normal. Later findings include low-frequency conductive hearing loss, dental crowding, and partial lack of secondary tooth eruption. Additional findings, present in some but not all affected individuals include photophobia, excessive ocular tearing, exposure keratitis, and Raynaud phenomenon. Cardiovascular disease leads to early death, as a result of complications from severe atherosclerosis, cardiac disease (myocardial infarction) or cerebrovascular disease (stroke). Average life span is approximately 13 years.

The diagnosis of this lethal disease is based on the recognition of common clinical features and the detection of either the classic c.1824C>T (p.Gly608Gly) heterozygous *LMNA* mutation or one of three heterozygous *LMNA* mutations in atypical HGPS: c.1822 G>A (p.Gly608Ser), c.1821 G>A (p.Val607Val), or c.1968+1G>A. The most prevalent mutation identified as a molecular origin of HGPS is the *LMNA* point mutation G608G where the nucleotide change activates a cryptic splice site leading to the deletion of 50 amino acids including the protease ZMPSTE24 proteolysis site. This truncated prelamin A, called Progerin (Δ 50 lamin A), remains farnesylated and strongly anchored in the nuclear membrane (Capeau et al., 2010). The exact mechanisms responsible for the premature aging process of HGPS are not fully understood. The gradual accumulation of progerin leads to serious defects in nuclear architecture and function (Goldman et al., 2004) (Capeau et al., 2010), preventing the recruitment of DNA repair elements (Dechat et al., 2008) and other regulatory proteins leading to genomic instability. Also *LMNA* mutations can cause an incomplete disassembly of nuclear envelope, chromosome missegregation and binucleation during mitosis (Cao et al., 2007; Dechat et al., 2008). Furthermore, progerin affects mesenchymal stem cells by altering differentiation of the cell lineages in a cell-specific manner (Scaffidi and Misteli, 2008). This observation is of seminal importance since the affected tissues in lipodistrophies are mainly of mesenchymal origin.

As an example, Scaffidi and Misteli (2008) described that differentiation of adipocytes from hMSCs was markedly reduced in hMSCs expressing progerin.

Progerin is also present in low levels in normal aging cells (Scaffidi and Misteli, 2006a). It is hypothesized that this presence of progerin is somehow triggered by shortened telomeres, and may therefore be associated with the irreversible process of cellular senescence (Cao et al., 2011). There is no doubt that new findings concerning laminopathies, in particular those caused by progerin production, will reveal determinants of the aging process. This is one of the reasons why the study of laminopathies is a highly active research area.

HIV-Related Lipodystrophy

HIV-related lipodystrophy is the most prevalent form of lipodystrophy. The prevalence of this disease varies depends on the different studies: between 20-80% of HIV infected people, hence approximately 5 to 10 million people may develop this disease (UNAIDS report 2012).

HIV-related lipodystrophy usually presents at clinics diagnostically with adipose tissue loss, especially subcutaneous (SAT) in the face, arms, buttocks, and legs; patients may also suffer fat deposits in the trunk and upper back and neck referred to as "buffalo hump". Breast size of both male and female patients tends to increase, patients also tend to develop abdominal obesity (Palchetti et al., 2013; Garg, 2011; Capeau et al. 2010). Fat loss, particularly in the face could lead to emaciated appearance with the stigmatization that this facial appearance entails. In addition, HIV-associated lipodystrophy also presents metabolic disorders, which are more severe than the aesthetic appearance, predominantly insulin resistance, hypertriglyceridemia, low HDL levels and hepatic steatosis which can leads to cirrhosis and liver insufficiency, (Capeau et al., 2010; Garg, 2011). In fact, these metabolic complications are responsible for age-related co-morbidities, such as increased cardiovascular and hepatic risks. Furthermore, HIV-infected patients present features of premature aging affecting bone, brain, vascular walls, muscles, kidney and liver. Although the mechanisms remain unclear the long term HIV infection, immune depletion and the adverse effects of some antiretrovirals are clearly involved (Caron-Debarle et al., 2010).

To assess the molecular pathways possibly involved numerous clinical trials have been designed which so far suggest several factors may be contributing.

Host factors, including low CD4+ cell counts (Lichtenstein et al., 2005), elevated levels of TNFα (Maher et al., 2002) and other inflammatory components (Maurin et al., 2005), are thought to play a role in development of lipodystrophy. Also, the subtype of infection (HIV-1) can itself contribute to the development of the lipodystrophic phenotype by activation of proinflamatory cytokine release and persistence of infected macrophages in adipose tissue of patients which have not received antiretroviral treatment (Crowe et al., 2010; Jacobson, 2005; Grinspoon and Carr, 2005; Caron-Debarle et al., 2010).

It is difficult to be certain about the contributions of a multitude of factors in the etiology of lipodystrophy, as it is a complex disease; however there is unequivocal evidence about the role played by medications used in the treatment of HIV infection. The development of lipodystrophy has been further associated with components of the antiretroviral therapy as specific nucleoside reverse transcriptase inhibitors (NTRI). NTRI are incorporated into the DNA of the virus resulting in an incomplete DNA that can not create a new virus. Mitochondrial toxicity and increased oxidative stress is postulated to be involved in the pathogenesis associated with NRTI and other factors (Caron et al., 2008; Murphy, 2004; Moyle, 2005; Gallant and Pham, 2003).

Since 1996, shortly after the introduction of protease inhibitors (PI) in antiretroviral therapy, it was reported that treated patients developed lipodystrophy (Grinspoon, S. and Carr, A. 2005). PIs are a fundamental feature of the "highly active antiretroviral therapy" (HAART), which has been demonstrated to improve survival of HIV infected patients. PIs block the HIV aspartyl protease, an essential enzyme for viral replication (Coffinier et al., 2007). However, some of these PIs also interfere with the processing of lamin A (Caron et al., 2003) by inhibiting the protease ZMPSTE24 (Coffinier et al., 2007). This inhibition leads to a significant accumulation of farnesyl-prelamin A relative to mature lamin A and the consequent metabolic/lipodystrophy syndrome associated as a side effect. Many experiments (*in vitro*, *ex vivo* and *in vivo*) have shown these drugs are toxic to adipose tissue and could act in synergy to produce complex clinical and biological alterations (Hammond and Nolan 2007; Caron et al., 2007b; Villarroya et al., 2010). Interference with lipid metabolism is postulated as a route to the pathophysiology of lipodystrophy.

At the same time, PI treatment could also be responsible for inhibiting differentiation of adypocytes and increasing proinflammatory cytokine release, inducing metabolic alterations and activating the NF-kB pathway (Capeau et al., 2010).

PI treatment could also affect adipose tissue by increasing oxidative stress and promoting premature cell senescence through inhibition of ZMPSTE24, and subsequent accumulation of farnesylated prelamin A (Caron et al., 2007a) (Caron-Debarle et al., 2010).

Moreover, Bastard and collaborators (Bastard et al., 2002) describe changes in mRNA levels, protein expression and morphology of major adipocyte differentiation markers as a result of treatment with PIs. For example, SREBP1c expression was significantly reduced in subcutaneous adipose tissue of HIV-1 infected patients who developed lipodystrophy while on protease inhibitor based HAART treatment. Given that the differentiation factor SREBP1 is rapidly targeted by protease inhibitors *in vitro*, their results *in vivo* suggest that SREBP1c could be an important mediator of peripheral lipoatrophy in this setting, leading to metabolic alterations such as insulin resistance. Subsequently Kratz and coworkers (Kratz et al., 2008) verified that decreased expression of genes involved in adipocyte differentiation, lipid uptake, and local cortisol production precedes HIV-associated lipodystrophy, in biopsies of subcutaneous adipose tissue of HIV patients undergoing HAART treatment.

In summary, mutations in *LMNA* or defective posttranslational processing of prelamin A (mutations in *ZMPSTE24*, f.e.) are responsible for more than ten different nuclear envelope disorders called laminopathies, which include systemic disorders and tissue restricted diseases (Capell and Collins, 2006; Verstraeten et al., 2007). Among them, the premature aging syndrome known as Hutchinson-Gilford Progeria syndrome (HGPS) and some partial lipodystrophies share a defective physiological maturation of prelamin A as the main pathophysiological mechanism. Progeria is caused by a *de novo* point mutation, which creates an efficient alternative splice donor site and interferes with conversion of farnesyl-prelamin A to mature lamin A. (De Sandre-Giovannoli et al., 2003). This leads to the production of a truncated lamin A protein with an internal deletion of 50 amino acids in the C-terminal domain of the protein. The missing sequence includes the proteolytic cleavage site recognized by ZMPSTE24. The mutant protein, known as progerin, is permanently farnesylated (Figure 2).

In the case of some partial lipodystrophies, such as acquired and some congenital (MADB) lipodystrophies, the activity of ZMPSTE24 is inhibited, leading the accumulation of farnesylated prelamin A (Coffinier et al., 2007) (Figure 2).

How and why prelamin A or progerin accumulation induces the lipodystrophy or the premature aging associated phenotypes remains unclear.

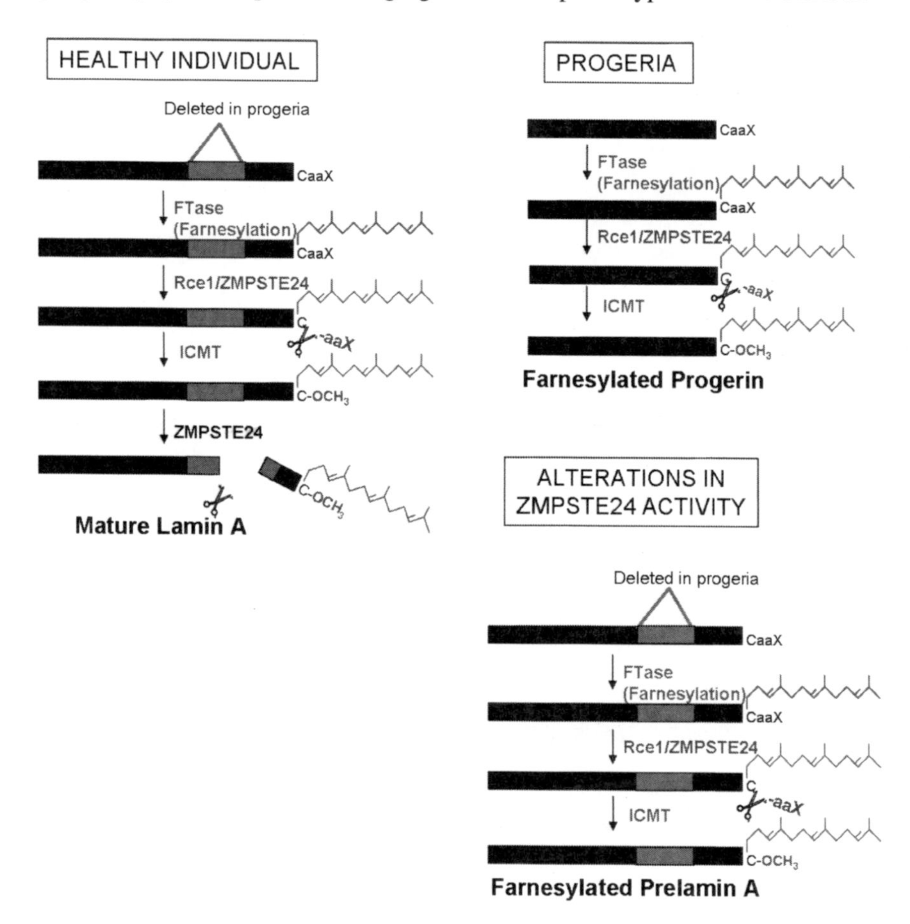

Figure 2. Maturation of Lamin A protein and formation of the permanently farnesylated Progerin and Prelamin A. The region deleted in progerin is highlighted in pink.

However, the generation of genetically modified animals to study these diseases has provided valuable clues for understanding the development of these laminopathies.

LMNA Murine Models

The modification of genes to analyze their functional roles has been extensively utilized for decades. One strategy for gene modification is homologous recombination in murine embryonic stem cells which has contributed extensively to the functional study of numerous genes in mammals by generating "knockout mice" of the gene of interest (Thomas and Capecchi, 1987). This technique, homologous recombination, has also been used to introduce point mutations into genes of interest, resulting in the expression of mutant protein products.

Animal models and especially murine models have played a key role in the understanding and study of several human diseases. Considerable progress in understanding how mutations in a single gene can produce such a broad spectrum of phenotypes is, in part, attributable to the abundant laminopathy mouse models that have been generated in recent years (Capell and Collins, 2006). In fact, animal models have been defined as excellent tools for identifying the molecular basis of disease and broadening knowledge about the physiopathology of these diseases in order to develop new therapeutic strategies (Benavides et al., 2001). These genetically modified mouse models have greatly aided in the characterization of the relationship between physiological functions of nuclear lamins and their contribution in the generation of laminopathies (Table 2).

The significant importance of A-type lamins in maintenance of nuclear architecture and gene expression (Broers et al., 2006; Verstraeten et al., 2007) has been studied using a lamin null mouse model. The A-type lamins were eliminated by gene targeting, removing exon 8, exon 9, exon 10 and part of exon 11 of *Lmna* (Sullivan et al., 1999).

As expected, mouse embryonic fibroblasts (MEFs) of *Lmna* null mice exhibited alterations in nuclear envelope integrity, such as changes in inner nuclear membrane morphology and aberrations in the heterochromatin layer underlying the nuclear envelope.

In addition, the delocalization of emerin in *Lmna*$^{-/-}$ MEFs mimics a molecular hallmark of Emery-Dreifuss muscular dystrophy (EDMD), a disorder characterized by retarded postnatal growth and premature death by muscular dystrophy and cardiomyopathy (Nikolova et al., 2004; Sullivan et al., 1999; Wehnert and Muntoni 1999).

Table 2. Mouse models of LMNA-lipodystrophies with their contribution

Model	Phenotype	Comments	Literature
Lmna$^{-/-}$ *or* lamin null mouse	Growth retardation, muscular dystrophy and cardiomyopathy with postnatal lethality. MEFs with nuclear envelope alterations and delocalization of emerin	Mouse manifests characteristics with Emery-Dreifuss muscular dystrophy (EDMD)	Sullivan (1999)
	Mouse does not show lipoatrophy or insulin resistance	Model does not resemble Familiar Partial Lipodystrophy Dunnigan (FPLD2) phenotype	Cutler (2002)
	Delocalization of emerin in *Lmna-/-* MEFs. Contractile dysfunction in lamin A/C deficient myocytes. Alterations in SREBP1 import	Model suggests defects in mechanical force transmission	Nikolova (2004)
G608G mice	No external characteristics of progeria. Progressive loss of vascular smooth muscle cells (VSMC) and extracellular deposition in older mice	The phenotype is limited to VSCM so it has been proposed that regions are more susceptible to mechanical stress	Varga (2006)
Lmna$^{L530P/L530P\ 1}$ *and* *Lmna*$^{\Delta 9/\Delta 9\ 1}$	Complete absence of the subcutaneous fat layer, severe growth retardation, degeneration of cardiac and skeletal muscle, abnormal dentition. Marked growth retardation and shortened lifespan of fibroblasts	An effort to create a mouse model of AD-EDMD resulted in a mouse model of progeria	Mounkes (2003)
	Same pathology associated to progeria. Defective *Wnt* signalling, affecting ECM synthesis	Establishment of functional link between lamin A and ECM	Hernández (2010)

Table 2. (Continued)

Model	Phenotype	Comments	Literature
Zmpste24$^{-/-}$	Growth retardation and premature death. Cardiomyopathy, muscular dystrophy and lipodystrophy	Identification of prelamin A as specific substrate for Zmpste24	Pendás (2002)
	Improvement of loss of subcutaneous fat and rib fractures	FTI could rescue some aging-like phenotypes	Fong (2006)
		Statins and amino-bisphosphanates lipodystrophy and aging	Varela (2008)
Zmpste24$^{-/-}$ *Lmna*$^{+/-}$	Diminished farnesylated prelamin A. MEFs showed reduced nuclear blebbing	Association of toxicity and farnesylated prelamin A accumulation	Fong (2004)
Lmna $^{HG/HG}$	Severe progeria-like disease phenotypes. Nuclear blebs in MEFs	First revelation of FTI effectiveness	Yang (2005)
	Loss of subcutaneous fat and rib fractures	FTI increased adipose tissue mass	Yang (2006)
		FTI improves survival	Yang (2008a)
Lmna $^{nHG/+}$	Expression of non-farnesylated progerin. Misshapen nuclei	Non-farnesylated progerin was less toxic than the farnesylated version	Yang (2008b)
Lmna $^{csmHG/+}$	Expression of non-farnesylated progerin. No disease phenotype	Structural changes in carboxyl of *Lmna* gene could be responsible of distinct severities in phenotype	Yang (2011)
Lmna$^{LAO/LAO}$	Production of lamin A directly, bypassing prelamin A. No disease phenotype	Suggestion of prelamin A processing is dispensable	Coffinier (2010)

Model	Phenotype	Comments	Literature
Lmna ^{G609G}	Farnesylated progerin accumulation by defective splicing of *Lmna* gene presents in HGPS patients. Main characteristics of aging-related syndromes	Demonstration that progerin is accumulated in physiological ageing. Use of antisense morpholino oligonucleotides as therapy for progeroid syndromes: targeting in defective splicing as a therapeutic approach	Osorio (2011)
FPLD mice	Inability to accumulate fat, age-dependant partial lipodystrophy , hepatic steatosis, and insulin resistance. Reduced levels of PPARγ mRNA	Mouse resembles the characteristics of human FPLD2 patients. Preadipocytes can not complete adipogenesis program	Wotjanik (2009)
Lmna $^{GT-/-}$	Growth retardation, skeleton muscle hypotrophy, decreased subcutaneous adipose tissue and metabolic derangements. Impaired adipogenic differentiation of MEFs	Critical effects of loss of lamin A/C on early postnatal development.	Kubben (2011)

MEF: mouse embryonic fibroblast; SREBP1: sterol regulatory element-binding protein; BAC: bacterial artificial chromosome; FTI: farnesyltransferase inhibitor; ECM: extracellular matrix; AD-EDMD: autosomal dominant Emery Dreifuss muscular dystrophy; PPARγ: peroxisome proliferator-activated receptor.

[1] They are the same mouse model with two different names (LmnaL530P/L530P: Mounkes 2003; LmnaΔ9/ Δ9 : Hernández 2010).

The Emery-Dreifuss muscular dystrophy is a laminopathy that affects striated muscle and it is caused by mutations in either the *LMNA* gene (autosomal dominant form, AD-EDMD) and in emerin (EDM), a nuclear envelope associated protein (X-linked form).

Dissimilar severities of the clinical manifestations including early contractures in tendons, progressive muscle wasting and weakness are observable and consistent with the type of heritage and the penetrance of the mutations (Bione et al., 1994; Bonne et al., 1999; Helbling-Leclerc et al., 2002).

In spite of the fact that the lamins are expressed in most adult mammalian tissues (Broers et al., 2006), both human and mouse EDMD phenotypes caused by mutations in lamin A/C and the deficiency of *lmna* gene respectively, are mainly restricted to muscular tissue (Helbling-Leclerc et al., 2002).

The identification of the mechanism/s responsible for the different laminopathy phenotypes is a highly active area of research.

One hypothesis was suggested based on a Hutchinson-Gilford Progeria syndrome mouse model, G608G mouse (Varga et al., 2006). The transgenic mouse was generated introducing a human bacterial artificial chromosome which contained the common G608G HGPS mutation, so this mouse line only expressed progerin from the *LMNA* gene. The mouse showed a progressive loss of vascular smooth muscle cells (VSMC) which were replaced by collagen and extracellular matrix, principally in large arteries. Having found that the stiffness of the cytoskeleton in lamin A/C deficient fibroblasts is lower than wild type cells when they are subjected to mechanical stress (Lammerding et al., 2004), it is reasonable that the severity of effects could depend on the intensity of the contractile forces exerted on each tissue. Thus, the compromised integrity of cardiac and striated muscle nuclear envelopes in mutant animals leads to more severe defects in large arteries than in other tissues since these locations are brought under maximal mechanical stress. Although the underlying molecular mechanisms are unclear, the model was presented as a tool for the study of vascular pathology in progeria (Varga et al., 2006). Another hypothesis proposes that the direct interaction of lamins with cytoskeletal proteins could be responsible for the pathogenesis of laminopathies. Desmin is an intermediate filament protein that links the nuclear surface to the cytoskeleton, enabling the transmission of contractile signals inside the cell. In the lamin null mouse model, the abnormal connections of the nuclear envelope to the desmin filament network and other cytoskeleton structures could result in the contractile dysfunction showed by lamin A/C deficient myocytes (Hernandez et al., 2010; Nikolova et al., 2004).

Likewise lamins interact with proteins involved in mechanical force transmission; thus it stands to reason that lamins could interact with other functional proteins such as gene transcription regulated proteins.

SREBP1 is a nuclear transcription factor which controls the expression of many proteins implicated in lipid homeostasis, adipogenensis and insulin sensitivity, among other functions (Bidault et al., 2011; Shao and Espenshade,

2012). In lamin A/C deficient mice, the increased proportion of cleaved SREBP1 in the cytoplasm and the subsequent reduced level in the nucleus when compared with wild type mice (Nikolova et al., 2004) suggested the possibility that alterations in the nuclear envelope could be involved in the development of lipodystrophies, possibly by defects in nuclear pore transport. Moreover, the discovery of accumulated prelamin A as a common feature of acquired and genetic lipodystrophies (Caron et al., 2007a) and the sequestration of SREBP1 by prelamin A in fibroblasts of mandibuloacral dysplasia (MAD) and familiar partial lipodystrophy Dunnigan type 2 (FPLD2) patients, (Capanni et al., 2005) suggested a role for SREBP1 in the physiopathology of these syndromes.

Overexpression of one SREBP1c isoforms (one of the three isoforms of SREBP, which are SREBP1a, SREBP1c, and SREBP2) in adipose tissue under the control of the adipocyte-specific aP2 promoter in mice surprisingly revealed a lipodystrophic phenotype with a reduction of white adipose tissue (WAT) and an increase in brown adipose tissue (BAT) (Shimomura et al., 1998). Overexpression of SREBP1 can result in reduced levels of mRNA adipocyte markers rather than increased lipogenic genes contrary to expectations, and a large fatty liver, severe insulin resistance and hyperglycemia, which were the main characteristics of the transgenic SREBP1 mouse. These observations resemble some, but not all, of the features of congenital generalized lipodystrophy (GCL) in humans (Seip and Trygstad, 1996). CGL is an autosomal recessive syndrome caused by mutations in several genes including *AGPAT2*, *BSCL2*, *CAV1* and *PTRF1*. The disorder is characterized by lack of white fat, insulin resistance, enlarged fatty liver and diabetes mellitus. The corresponding mouse model exhibited an incomplete loss of interscapular fat pads, highlighting differences in the dissimilarity of WAT and in adipose tissue biology among humans and mouse.

Before death, *lmna* null mice display a deficiency of fat, suggesting this mouse might serve as a model for Dunnigan´s Familiar Partial Lipodystrophy (FPLD2). As has been previously mentioned in the classification of *LMNA*-linked lipodystrophies, FPLD2 is an autosomal dominant disease characterized by progressive regional subcutaneous fat loss (Seip and Trygstad, 1996) beginning at puberty and insulin resistance, which often progresses to diabetes (Dunnigan et al., 1974). In most cases, this syndrome is caused by missense mutations located in the globular C-terminal domain of the lamin A/C protein. *Lmna*$^{-/-}$ mice had severe muscular dystrophy and a lack of fat. But contrary to expectations, *Lmna* null mice did not show altered serum glucose or insulin levels as observed in FPLD2 patients with insulin resistance. Taken together these observations suggested that the adipose deficiency of homozygous *Lmna*$^{-/-}$ mice was due to the muscular dystrophy and the subsequent inability to feed, rather than due to FPLD2 pathophysiology (Cutler et al., 2002).

One of the most frequent mutations causing EDMD in humans is a substitution of proline for leucine at residue 530 of the *Lmna* gene. Surprisingly, the phenotypes associated with this point mutation in a *Lmna* knock-in mouse were consistent with those observed in progeria patients. Specifically, there is a complete absence of subcutaneous fat, hypoplasia and degeneration of cardiac and skeletal muscle, osteoporosis and abnormal dentition. Notably, the hallmarks of progeria, marked growth retardation and shortened lifespan, were also observed in mutant fibroblasts from these *Lmna*$^{\Delta9/\Delta9}$ mice (Mounkes et al., 2003). Later studies with the *Lmna*$^{L530P/L530P}$ mouse demonstrated that the truncated and unprocessed form of lamin A present in mutant fibroblasts remained farnesylated and changes in the expression of extracellular matrix (ECM) genes were observed which revealed an inhibition of *Wnt* pathway signaling resulting in defective ECM synthesis (Hernandez et al., 2010). The mouse embryonic fibroblasts (MEFs) were not able to proliferate adequately due to their dysfunctional ECM, revealing a link between nuclear lamins, ECM and tissue proliferation. ECM undergoes a dynamic remodeling during adipogenesis (Mariman and Wang, 2010). Intriguingly, the tissues most acutely affected both in progeria patients and *Lmna*$^{L530P/L530P}$ mouse have mesenchymal origin, suggesting either an impairment in the differentiation of this cell type in both conditions, or an activation of an abnormal transdifferentiation program during old age (Mounkes et al., 2003).

Development and Testing of Potential Therapeutic Strategies: The Prenylation Dilemma

According to the lipodystrophic phenotype related to the alteration of prelamin A posttranslational processing, several mouse models have been created. The first model was developed deleting *Zmpste24* allele through gene targeting which encodes the metalloproteinase that cleavages immature form of Lamin A/C (Pendas et al., 2002). The increased nuclear abnormalities in *Zmpste24*$^{-/-}$ MEFs suggested that the absence of *Zmpste24* produces the compromised integrity of nuclear envelope. The *Zmpste24* null mice manifested cardiomyopathy, muscular dystrophy and lipodystrophy, among other characteristics, similar to those observed in human laminopathies patients (Bonne et al., 1999; Fatkin et al., 1999; Shackleton et al., 2000). The *Zmpste24*$^{-/-}$ mouse model demonstrated that prelamin A was a specific substrate for Zmpste24 and the necessity of this enzyme for the *in vivo* prelamin A posttranslational modification (Pendás et al., 2002). In order to clarify whether the accumulation of non-processed forms of lamin A and/or prelamin A could be toxic in *Zmpste24*$^{-/-}$ mice, farnesylated prelamin A levels were reduced through the elimination of one *lmna* allele (Fong et al., 2004). These *Zmpste24*$^{-/-}$ *lmna*$^{+/-}$ mice were normal and they did not show any disease phenotypes, such as nuclear blebbing, suggesting that farnesylated prelamin A was toxic and the reduction of its levels *in vivo* improves the aging related phenotypes observable in the HGPS mouse model.

Supporting the hypothesis that irreversible farnesylation of prelamin A is responsible for the toxic effects evidenced by its accumulation in aging-like phenotypes due to its targeting to the nuclear rim (Fong et al., 2004), a mouse model that produce only progerin from the *Lmna* gene was developed (Yang et al., 2005). This HGPS model was created through a gene-targeting strategy which consisted of a deletion of intron 10 (which abrogates lamin C synthesis) and a deletion of intron 11 and the last 150 nt of exon 11 of a *lmna*$^{HG/HG}$ allele. The treatment with a farnesyltransferase inhibitor (FTI), which prevents the farnesylation step of the Lamin A processing, showed a diminished number of *lmna*$^{HG/+}$ MEFs with nuclear blebs compared to *lmna*$^{HG/HG}$ MEFs without treatment. Hence, due to the farnesyltransferase inhibitor treatment, the decreased mislocalization of progerin at the nuclear rim produced a reduction in the cellular phenotypes of aging (Yang et al., 2006).

Moreover, the treatment of *Zmpste24*$^{-/-}$ mice (as another established model of HGPS) with FTI improves the loss of body weight and incidence of rib fractures normally seen in these mutant mice (Fong et al., 2006). Recent findings have also demonstrated that FTI extends longevity in *lmna*$^{HG/+}$ and *Zmpste24*$^{-/-}$ mice as well (Yang et al., 2008a) (Fong et al., 2006).

Along with to the improvement of phenotypes by FTI treatment in mouse models of progeria listed above, the number of spontaneous rib fractures and body weight were ameliorated in the transgenic mouse that showed cardiovascular phenotype characteristic of progeria (Capell and Collins, 2006; Varga et al., 2006). But the disease phenotype was not completely rescued with the FTI treatment in all cases. To investigate the importance of farnesylation in the development of disease phenotype, a genetically identical model of *lmna*$^{HG/+}$ mouse replacing a cysteine with a serine that prevents farnesylation of progerin designated the *Lmna*$^{nHG/+}$ mouse was generated (Yang et al., 2008b). Taking the misshapen nuclei as a measure of the severity of the phenotype, the *Lmna*$^{nHG/+}$ mouse exhibited milder phenotypes than the *Lmna*$^{HG/+}$ mouse, suggesting that non-farnesylated progerin was less toxic than the farnesylated version. However, the mild phenotype exhibited by *Lmna*$^{nHG/+}$ mouse and FTI-treated *Lmna* $^{HG/+}$ mouse highlighted the farnesylation was not the unique factor involved in the physiopathology of these syndromes (Yang et al., 2008a).

Another mouse model that yields non farnesylated progerin was developed through the in-frame deletion of three nucleotides of the carboxy terminal motif which abolished the protein prenylation, the *Lmna*$^{csmHG/+}$ mouse (Yang et al., 2011). Even though levels of progerin were similar those found in the first mouse model that produces non farnesylated progerin (*Lmna*$^{nHG/+}$ mouse), no disease phenotype was observed. Taken together, these studies suggest that minimal changes in carboxy terminus of progerin would cause huge differences in toxicity of the mutant protein which would be responsible of the several grades of severity in disease phenotypes (Yang et al., 2008a; Yang et al., 2005; Yang et al., 2011; Yang et al., 2008b).

The absence of a complete rescue of the phenotype as a consequence of the FTI treatment could be due to alternative prenylation of prelamin A as has been previously described for oncoproteins such as K-Ras (Whyte et al., 1997). Moreover, the possibility of an alternative prenylation of the carboxyl terminus by geranylgeranyltransferases (GGTases) suggests an explanation for the limited efficacy of FTI treatment (Fong et al., 2006; Yang et al., 2005; Yang et al., 2008b). With the aim of assessing this hypothesis, *Zmpste24*$^{-/-}$ mouse was treated with a combination of statins and amino-bisphosphonates (Varela et al., 2008). Statins and bisphosphonates are compounds that avoid the synthesis of isoprenyl lipids which are necessary to successfully complete the maturation of lamin A. The aging-like phenotype of the *Zmpste24*$^{-/-}$ mouse (such as lipodystrophy measured as amount of subcutaneous fat) was ameliorated by the treatment, broadening the perspectives for therapy strategies for lipodystrophies and aging diseases related to *LMNA* gene and its translational changes (Varela et al., 2008).

All these studies pointed to the accumulation of prelamin A as responsible for the lipodystrophic phenotypes. Another strategy for treatment would be to decrease the accumulation of prelamin A by directly synthesizing mature lamin A, as long as prelamin A is not required for another purpose to the cells. To address this question, a mouse model was developed which produces lamin A directly, bypassing prelamin A synthesis and processing (Coffinier et al., 2010). The allele generated by gene targeting, *Lmna*^{LAO} ("mature lamin A-only") contains a stop codon immediately after the last codon of mature lamin

A. This manipulation results in the abolishment of lamin C synthesis entirely, and concomitantly produces an overexpression of mature lamin A without any production of lamin C or prelamin A in the homozygous mice. In this mouse, body weight was normal, and no disease phenotype was detected except for an increased frequency of nuclear blebbing.

Moreover, the subcellular localization of mature lamin A in homozygous mice was at the nuclear rim, with no observable differences compared to wild type mice. These results demonstrate that prelamin A processing is dispensable, at least in mice.

Another approach is to target the abnormal *LMNA* splicing. The possibility of modulation of progerin splicing to control the transcript level of prelamin A led to development of a knock-in mouse that carried the *Lmna*^{C609G} allele, the same mutation that produces a defective splicing of *Lmna* gene in HGPS patients, which results in accumulation of farnesylated progerin (Osorio et al., 2011). This mouse accumulated progerin and manifested the main characteristics of aging-related syndromes resemble HGPS. The defective splicing that yields progerin in these knock-in mice was modulated owing to an approach based on morpholinos. Morpholinos are small oligonucleotides that avoid the access of ribonucleoproteins to donor and acceptor sites, altering RNA splicing efficiency (Parra et al., 2011). Hence, progerin levels were diminished to undetectable amounts in *Lmna*^{C609G} mice and the symptoms *in vivo* were ameliorated, opening a new hopeful therapeutic opportunity with these synthetic reagents (Osorio et al., 2011).

Challenging Mouse to Model Human Lipodystrophy

Discrepancies between mouse and humans previously mentioned in this chapter prompt us to question into the suitability of these animals to study human lipodystrophies.

The "lipid overflow" hypothesis proposes that the capacity of adipose tissue to accumulate lipids is finite, and when adipocytes reach their maximum capacity to store lipids, excess lipid is stored in ectopic regions (Danforth, 2000; Savage, 2009). PPARγ is a well-known marker of adipocyte differentiation, and many studies have related these fat depot specific differences with changes in PPARγ mRNA expression (Tchkonia et al., 2002). Reduced levels of PPARγ in mutant murine FPLD2 cells (Wojtanik et al., 2009) suggest that preadipocytes from FPLD2 mice are incapable of developing into mature adipocytes and accumulating lipid (Savage, 2009).

Moreover, the hypothesis that the pool of adipocytes is finite supports the same idea. If adipocytes complete the adipogenesis program slower than is necessary to the normal turnover of adipose tissue, there will not be enough fat precursors for correct lipid accumulation. These hypotheses were arrived at through two mouse models. First, a mouse model of generalized lipodystrophy (Moitra et al., 1998) was created, wherein the transplantation of wild type mouse fat improved metabolic complications in the mutant mouse. Second, an inducible transgenic mouse model of acquired lipodystrophy was created through ablation of fat tissue by a chemical method.

Mice were treated with a synthetic compound which forces the dimerization of caspase 8, resulting in an "inducible fatless mouse" (Pajvani et al., 2005). The main metabolic alterations showed by the mice were a marked reduction of adipokines, such as adiponectin, resistin and leptin, leading to glucose intolerance and hyperphagia. After chemical treatment finalized, adipocytes recovered their functional properties. The reversibility of the fat loss allows for the latter mouse model to be used as a tool to study adipogenesis process *in vivo*. These insights together support the idea that the lack of fat contributes to the lipodystrophic phenotype besides demonstrating that mice can develop lipodytrophy (Gavrilova et al., 2000).

During development lamins seems to be regulated in a similar way, since both in mice and humans A-type lamins are developmentally regulated in a tissue-dependent manner. In mice, lamin A/C is not expressed before day 9 of development, and in humans its expression is activated during human embryonic stem cell (hESC) differentiation, suggesting that lamins A and C are involved in the maintenance of the differentiated state (Constantinescu et al., 2006; Gonzalez et al., 2011; Stewart and Burke, 1987). Several *LMNA* mutations affecting cell differentiation have been studied, but only in young adult mice.

To extent knowledge about the role of lamins in early postnatal development, a *Lmna* deficient mouse with a *Lmna* gene disrupted by a reporter gene was created, $Lmna^{GT-/-}$. When the *lmna* gene promoter was activated during normal development, reporter activity was visualized on day 11 of embryonic development (Kubben et al., 2011). *Lmna* transcripts were detected in several tissues in early embryonic development and their absence produced severe growth retardation, decreased subcutaneous tissue deposits and heart defects causing death associated with muscle weakness and metabolic complications at 2 weeks. Additionally, mice presented altered metabolic parameters. Mutant mouse embryonic fibroblasts (MEF) induced to carry out an adipogenesis program *ex vivo* for 3 weeks, showed impaired adipogenic differentiation. Overall, these facts demonstrated the critical role that lamin A/C play in the successful development of animal tissues.

Focusing on *LMNA*-linked lipodystrophy murine models, the missense mutation at arginine 482 of exon 8 of *LMNA* gene attracts the attention, since this mutation accounts for more than 85% of FPLD2 cases in humans. A transgenic mouse which specifically expressed either the human mutant lamin *A* or *C* containing this R482Q FPLD2 mutation in adipocytes was generated (Wojtanik et al., 2009).

Consistent with FPLD2 patients, the transgenic mouse was incapable of accumulating fat to the same extent as its wild type counterpart, and the mutant mice manifested age-dependent partial lipodystrophy, hepatic steatosis and insulin resistance, but with several differences compared to humans. In humans, the lipodystrophic phenotype and metabolic complications trigger at the time of puberty, which is significantly earlier than the onset observed in mutant mice. Also visible FPLD2 phenotypes are more frequent in women than in men contrary to the mice suggesting a sexual dimorphism in humans that is not apparent in the mouse. Altogether, these data suggest that at least some of the modifications in lipodystrophy between species are influenced by hormonal control (Savage, 2009; Wojtanik et al., 2009). Furthermore, differences have been found in the pattern of lipodystrophy. All fat pads are smaller in transgenic mutant mice, whereas humans with FPLD have an excess of neck and facial fat, but lose femorogluteal, limb and truncal fat (Hegele, 2001). While lipolysis of adipocytes in mutant mice is normal, the mice seem to have difficulties in differentiating and accumulating adipocytes (Hegele, 2001; Wojtanik et al., 2009).

On the other hand, lipodystrophic syndrome (acquired lipodystrophy) has also been associated with the use of human immunodeficiency virus (HIV) protease inhibitors (PIs), as a part of the HAART-linked lipodystrophies (high active antiretroviral therapy). After the discovery that HIV protease inhibitors modify the subcellular localization of SREBP1 in mouse preadipocytes and mesenchymal stem cell lines, coinciding with the impairment of insulin signalling and adipogenic differentiation in variable degrees (Caron et al., 2003; Caron et al., 2001; Dowell et al., 2000; Lenhard et al., 2000; Vernochet et al., 2003), numerous studies have explored the probable underlying mechanisms. Moreover, it has been established that PIs, at physiologically relevant concentrations, cause substantial accumulation of prelamin A, by specifically inhibiting ZMPSTE24 activity *in vitro* (Coffinier et al., 2007).

The pathogenesis of HAART-linked lipodystrophies has also been studied at the transcriptional level. Microarray studies have identified new genes modulated by PIs and nucleoside reverse transcriptase inhibitors (NRTIs), in 3T3-L1 mouse preadipocytes under adipogenic conditions (Pacenti et al., 2006). NRTIs are another antiretroviral drug component of the long-term HIV therapy, HAART. Both PI and NRTI inhibited adipocyte differentiation *in vitro*, with NRTI´s showing milder effects than PIs. Master adipogenic transcription factors were modulated by PI, such as cebpα and pparγ, as well as adipocyte-specific markers (e.g. adiponectin, leptin and Cd36), suggesting that altered gene expression may be one of several molecular mechanisms implied in HAART-linked lipodystrophies (Pacenti et al., 2006).

Despite the significant increase in the knowledge of the mechanism and course of human diseases provided by murine models, both *in vivo* and *in vitro*, there are downsides in the use of these lipodystrophic models. One such handicap is the generation of reliable mouse models of partial lipodystrophy. For example, in many *LMNA*-linked disorders it has been observed that heterozygous mutant mice had milder phenotypes than those detected in human heterozygous patients. Furthermore, in many instances the homozygous mouse is a significantly better model for the study of the parallel human autosomal dominant syndrome (Navarro et al., 2005; Osorio et al., 2011; Pendás et al., 2002).

In addition several studies highlight differences in gene expression between humans and mice (Svensson et al., 2011) both in white adipose tissue (WAT) and in brown adipose tissue (BAT). Also, the distribution of adrenergic receptors in fat tissue is different, so the functional responsiveness between species is variable (Lafontan and Berlan, 1993; Mattsson et al., 2011).

With regard to the transcriptional regulation of the *lmna* gene, additional promoter cell-germ specific regulatory elements have been identified in mice which have not been identified in humans (Furukawa et al., 1994; Nakajima and Abe, 1995; Dechat et al., 2008; Broers et al., 2006).

Another example of differences between mouse and human species regarding the study of laminopathies is the different phenotypes obtained by the same mutation: In spite of the high homology known between the human and mouse lamin A gene, the same point mutation, L530P produces an aberrant splicing of lamin in mouse (Mounkes et al., 2003) which has still not been demonstrated in human patients. The resultant lmna transcripts are unstable and they result in progeria. In contrast the human L530P mutation causes AD-EDMD in humans (Bonne et al., 1999).

All these interspecies differences underscore the necessity of human models, both to unravel the physiopathological mechanisms involved in

LMNA-linked lipodystrophies and to discover putative targets for therapeutic compounds.

There is a need for knowledge about the role of lamin A in the regulation of signaling pathways and how these altered signals contribute to the development of lipodystrophies. Why prelamin A or progerin accumulation induces the lipodystrophy or the premature aging associated phenotypes remains unclear. Since *LMNA*-linked lipodystrophic syndromes and progeria primarily affect tissues from mesenchymal origin such as the adipose lineage, one hypothesis is that prelamin A accumulation in these cells hampers the proper process of adipogenesis from mesenchymal stem cells (Ruiz de Eguino et al., 2012). This indicates that mesenchymal stem cells could be playing a relevant role in the pathophysiology of *LMNA*-linked lipodistrophies.

Adipocyte Differentiation from Mesenchymal Stem Cells

Adipogenesis is the biological process by which new adipose tissue is generated. Adipose tissue plays major roles in energy homeostasis, lipid metabolism and insulin actions. In mammals, two types of adipose tissue, with different morphological features and functions, have been described: white adipose tissue (WAT) and brown adipose tissue (BAT) (Cinti, 2005). The main role of WAT is the regulation of energy homeostasis through the storage and release of triacylglycerol (Ailhaud et al., 1992; Rosen and Spiegelman, 2006). However, in addition to being a fat reservoir, WAT is also an endocrine organ since white adipocytes secrete adipokines involved in the regulation of food intake and energy homeostasis, as well as in the inflammatory states associated with metabolic disorders (Rosen and Spiegelman, 2006; Waki and Tontonoz, 2007). BAT, on the other hand, has a thermogenic activity, regulating body temperature by dissipating energy through heat production (Spiegelman and Flier, 2001). As a matter of fact, this is the primary thermogenic mechanism for small mammals and human neonates to prevent hypothermia (Klingenspor, 2003). BAT uses the chemical energy in lipids and glucose to produce heat due to high mitochondrial content and the ability to uncouple cellular respiration through the action of uncoupling protein-1 (UCP-1). UCP-1 is localized to the inner mitochondrial membrane and acts to uncouple oxidative phosphorylation from ATP production, thereby releasing energy as heat (thermogenesis) (Cannon et al., 2004). Thus, while white adipocytes are characterized by a single, large lipid droplet and few mitochondria, brown adipocytes contain several small lipid droplets and many mitochondria. Importantly, recent studies have indicated that in addition to thermogenesis, BAT is involved in triglyceride clearance and glucose disposal (Bartelt et al., 2011) as well as the secretion of BAT-derived adipokines known as BATokines (Chartoumpekis et al., 2011; Hondares et al., 2010).

Although most WAT development occurs during late prenatal and early postnatal life, WAT can expand during adult life when energy intake exceeds energy expenditure. On the contrary, it was assumed for a long time that BAT is only present in newborns or infants and absent or very scarce in adults so that its contribution to energy expenditure was considered irrelevant. However, the use of non invasive techniques such as PET/CT-scans (Positron Emission Tomography associated with Computed Tomography) in recent years has demonstrated that BAT is indeed present in healthy adult individuals. These depots are mainly observed in the supraclavicular, suprarenal, paravertebral regions and in the neck (Nedergaard et al., 2007; Cypess et al., 2009; Zingaretti et al., 2009).

Mesenchymal stem cells (MSCs) are undifferentiated multipotent stem cells that reside in almost all tissues such as bone marrow, umbilical cord and adipose tissue (Minguell et al., 2001). These cells can be isolated easily, and they possess a high proliferative and differentiation capacity to differentiate (both *in vitro* and *in vivo*) into several mesenchymal lineages such as fat, cartilage and bone (Pittenger et al., 1999). In adult organisms, MSCs are the physiological progenitors of adipocytes (Park et al., 2008a). On the other hand, white and brown adipocytes have long been assumed to share a common developmental origin because they express a common set of genes involved in triglyceride metabolism and they undergo a very similar program of differentiation (Rosen and Spiegelman, 2000; Hansen and Kristiansen 2006; Gesta et al., 2007). Later studies, however, have indicated that brown and white fat cells, in fact, arise from distinct cellular lineages. Therefore, precursors of brown fat cells express the myogenic transcription factor Myf5 (Seale et al., 2008), a gene previously assumed to be present almost exclusively in committed skeletal muscle precursors. In contrast, progenitors of white adipocytes do not express Myf5 (Figure 3). Thus, during the developmental process, MSCs give rise to distinct subtypes of adipocyte precursors (Myf5- and Myf5+) that differentiate into white or brown adipocytes. Nevertheless, caution must be taken with a strict division between My5- and Myf5+ progenitors, since a very recent publication describes subsets of white adipocytes originating from both My5- and Myf5+ progenitors

(Sanchez-Gurmaches et al., 2012). Importantly, recent studies in mice have described a type of adipocyte that share characteristics of both white and brown adipocytes, the so-called brite (white + brown) or beige adipocyte (Seale at al., 2008; Barbatelli et al., 2010; Petrovic et al., 2010). These brite/beige adipocytes, found in some classical white fat depots, possess many of the morphological characteristics of brown adipocytes, but they do not express myocyte-enriched genes, indicating that they arise from a non-Myf5 cell lineage.

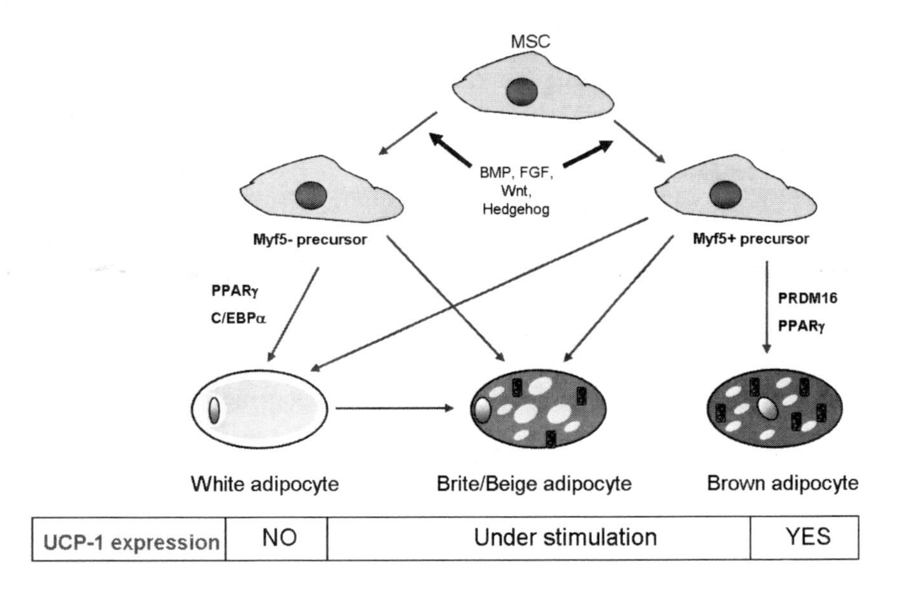

UCP-1 expression	NO	Under stimulation	YES

Figure 3. Overview of subtypes of adipocytes and their origins. Three types of adipocytes can be distinguished, the classic brown and white adipocytes and the brite adipocytes. Ucp-1 expression can be considered as a marker of brown adipocytes as well as brite adipocytes.

Moreover, these cells posses brown adipocytes functional characteristics. Although in the basal state they express very little UCP1, once stimulated (with cold exposure, for example) they activate expression of UCP1 to levels that are similar to those of the classic brown fat cells (Wu et al., 2012). More strikingly, a recent report described that brown fat depots in adult humans exhibit a molecular signature that is more comparable to that of murine brite/ beige adipocytes (Wu et al., 2012). This work supports the hypothesis that the BAT in mice present at birth is different from the thermogenic adipose tissue observed in human adults. Regarding the developmental origin of these cells, whether these beige cells derive from existing stem cells or from the direct transdifferentiation of differentiated white adipocytes or from a combination of these phenomena, is still not fully understood.

The development and differentiation of distinct adipocyte lineages is driven by a set of highly coordinated transcriptional events that reprograms the genome and turns on a set of genes characteristic of the specialized adipocyte. Although adipocyte formation is a multistep process involving many cellular intermediates, it is typically described in the context of two major phases: the "determination" and "terminal differentiation" stages.

In the determination phase, mesenchymal stem cells become committed to the adipocyte lineage and lose their ability to differentiate into other mesenchymal lineages. In this phase, committed preadipocytes are morphologically indistinguishable from their precursors. Subsequently, during terminal differentiation, the committed cells differentiate and acquire the phenotypical and molecular characteristics of mature adipocytes (Fevè, 2005). Functionally, adipogenesis reflects a fundamental shift in gene expression patterns within uncommitted MSCs that promotes and culminates in the phenotypic properties that define mature adipocytes. Despite the differences in the developmental origins and physiological functions of white and brown adipocytes, both cell types share a very similar transcriptional cascade that controls the process of fat differentiation. The adipogenic program of both white and brown adipocytes is based on the expression and activation of PPARγ (peroxisome proliferator-activated receptor-γ), the lipid–activated nuclear hormone receptor which plays the role of the master transcriptional regulator of adipogenesis (Rosen and Spiegelman; 2000). Moreover, this transcription factor is not only crucial for adipogenesis but is also required for the maintenance of the differentiated state (Tamori et al., 2002; Imai et al., 2004). C/EBPs (CCAAT/enhancer binding proteins) (C/EBPα, C/EBPβ, C/EBPsδ) function cooperatively with PPARγ and induce a transcriptional cascade that promotes and maintains the stable differentiated state of adipocytes (Rosen and Spiegelman 2000; Hansen and Kristiansen, 2006; Gesta et al., 2007). Among them, C/EBPα is essential for the normal insulin sensitivity of mature fat cells, and it is required for the formation of white cells, but not brown cells. C/EBPβ and C/EBPδ, as well as other transcription factors, also participate in the transcriptional cascade of adipogenesis by regulating PPARγ gene expression. Brown fat cell differentiation requires PPARγ, but this factor alone is not sufficient to drive MSCs into a brown fat program. In fact, the interaction of C/EBPβ with the zinc finger-containing protein PRDM16 is required, which leads to adipogenic development through induction of mitochondrial biogenesis via subsequent activation of peroxisome proliferator-activated receptor γ-coactivator 1α (PGC-1α) (Uldry et al., 2006). Importantly, brite/beige adipocytes express a unique gene expression profile that is distinct from either white or brown adipocytes (Wu et al., 2012). Thus, the beige cells express very little of the thermogenic program, including UCP1 in the basal state. However, once stimulated, these beige cells activate the expression of UCP1 to levels that are similar to those observed in brown cells (Figure 3).

Human Adipogenesis Cell Models Based on Stem Cells

Although WAT is easily obtained from patients through biopsy procedures, it is difficult to maintain and cannot be expanded in culture. To overcome these obstacles, there are human white adipocytes cells lines obtained from biopsies of spontaneous tumors of WAT such as liposarcomas (Wabitsch et al., 2000; Hugo et al., 2006) or immortalized from human WAT (Darimont et al., 2003). In contrast, human BAT is unavailable unless highly invasive surgery is performed; there is only one immortalized human BAT – derived cell line (Zilberfarb et al., 1997; Kazantzis et al., 2012). Thus, in spite of their usefulness for the study of human adipogenesis, these WAT and BAT cell lines may not accurately reflect the physiological adipogenic process due to their transformed phenotype. Regarding murine models, there are *in vivo* discrepancies between mouse and human brown and white adipose tissue function and differentiation, making mouse cellular models of adipogenesis difficult to analyze (Himms-Hagen., 1999; Lafontan and Berla, 1993; Svensson et al., 2011). Moreover, recent studies have shown clear differences between rodent and human mesenchymal stem cells with respect to certain signalling pathways involved in adipocyte differentiation, such as the Hedgehog and BMPs pathways (Fontaine et al., 2008; Svensson et al., 2011; Pisani et al., 2011). For these reasons the use of primary human cells are crucial for trying to understand the physiological mechanisms regulating adipogenesis.

Due to the rarity of lipodystrophies, access to patient samples is difficult. Although much of the information related to this disease has come from cultured patient fibroblasts (Caron et al., 2007a) or from established cells lines and murine models expressing mutated human *LMNA* transgenes, none of these models recapitulate all of the phenotypes seen in the human disease (Capanni et al., 2005) (Scaffidi and Misteli, 2008; Hernandez et al., 2010). It is clear that either the murine origin of models or the non-adipocyte cells used in these experiments could explain these discrepancies.

Recently a number of human disease models have been established in order to overcome this deficiency. These disease models are based on MSCs or on induced pluripotent stem cells (iPSCs), and take advantage of the capacity of these cells to differentiate to certain cell types which are affected in *LMNA*-linked lipodystrophies, such as adipocytes. Moreover, these cells are capable of differentiating to functional adipocytes (Elabd et al., 2009; Ahfeldt et al., 2012), white adipocytes (Rodriguez et al., 2004; Morganstein et al., 2010; Ahfeldt et al., 2012), and more recently, into brite adipocytes (Elabd et al., 2009; Pisani et al., 2011).

A Model of Lipodystrophy Based on Prelamin a Accumulating hMSCs

Since MSCs are the physiological progenitors of adipocytes in adult organisms (Park et al., 2008) they represent a feasible model for studying the adipogenic process from the earliest phases of differentiation to a more mature phenotype. In recent years, a number of groups have developed human-cell-based models for the study of adipogenesis using hMSCs from different sources (Jain and Lenhard, 2002; Mackay et al., 2006; Elabd et al., 2009). These hMSC-based cell models have been demonstrated to differentiate towards both white and brown adipocytes (Pittenger et al., 1999; Elabd et al., 2009) and even into brite adipocytes (Pisani et al., 2011) under specific culture conditions.

A model of *LMNA*-linked lipodystophy based on hMSCs has been generated in our laboratory (Ruiz de Eguino et al., 2012) to study the patho-molecular mechanisms underlying prelamin A accumulation in hMSCs. The lipodystrophy model in hMSCs was generated with the use of an HIV protease inhibitor (PI) that had been previously demonstrated to inhibit ZMPSTE24 activity in human fibroblasts (Coffinier et al., 2007) resulting in prelamin A accumulation (Figure 4). Since both the genetic and acquired *LMNA*-linked lipodystrophies mainly affect tissues from mesenchymal origin, we hypothesize that prelamin A accumulation in hMSCs could compromise their homeostasis, preventing the correct adipogenesis process. Supporting this hypothesis, previous results showed that signaling pathways required for maintaining normal stem cell function were altered in hMSCs expressing progerin and in *Zmpste24*-null progeroid mice (Scaffidi and Misteli, 2008; Espada et al., 2008).

We observed that hMSCs treatment with PI resulted in an inhibition of adipogenesis, as a consequence of prelamin A accumulation. In fact, we showed for the first time the induction of the accumulation of prelamin A by PIs in hMSCs *in vitro*. This hMSCs-based experimental model recapitulated the features previously observed in fibroblasts treated with PIs, as well as in *LMNA*-linked lipodystrophies, such as a reduction in their proliferation capacity, increased cellular senescence and significant percentages of nuclei with altered heterochromatin distribution.

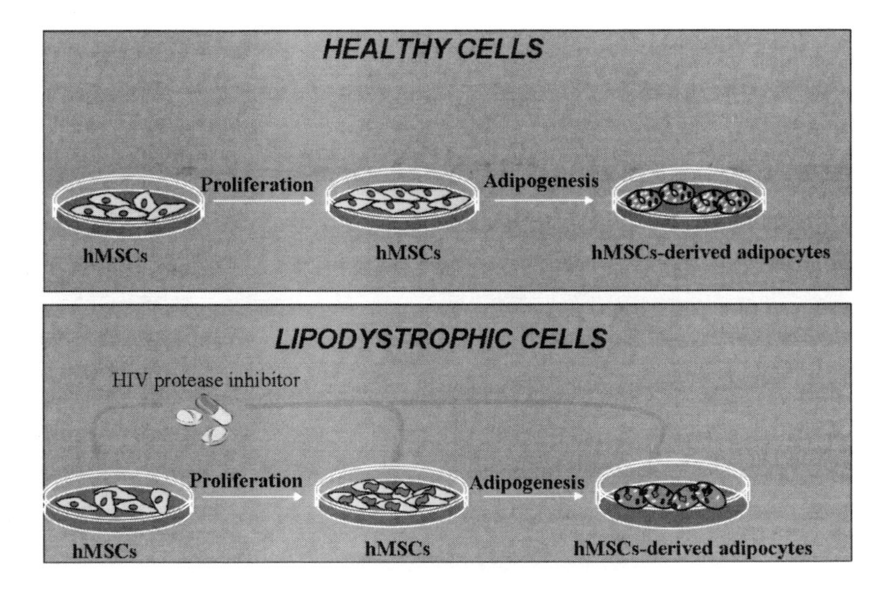

Figure 4. Schematic diagram of the lipodystrophy cell model based on hMSCs.

These observations validate PI-treated hMSCs as an experimental model for the study of *LMNA*-linked lipodystrophy.

Given that nuclear lamins regulate gene expression in cells by their association with either chromatin or transcription factors (Broers et al., 2006; Verstraeten et al., 2007), it can be hypothesized that the accumulation of prelamin A could result in sequestering transcription factors in hMSCs impairing their function as lipid metabolism regulators and thus leading to the lipodystrophy observed in these cells (Figure 5).

Analysis of the transcriptome profile of the experimental model of *LMNA*-linked lipodystrophy hMSCs showed alteration in expression of genes involved in lipid homeostasis, but surprisingly, also in genes encoding extracellular matrix proteins. Furthermore, in silico analysis suggested that the transcription factor Sp1, which regulates extracellular matrix genes, could be a potential regulator of the dysregulated genes altered in the lipodystrophy model. Indeed, Sp1 and prelamin A were physically interacting in adipocytes derived from hMSCs treated with PIs. Moreover, this interaction decreased Sp1 transcriptional activity in the prelamin A accumulated hMSC derived adipocytes. These results are supported by a previous publication (Capanni et al., 2005) in which the authors described a pathological sequestration of the transcription factor SREBP1 by prelamin A at the nuclear envelope in MAD and FPLD patient fibroblasts and in murine pre-adipocytes.

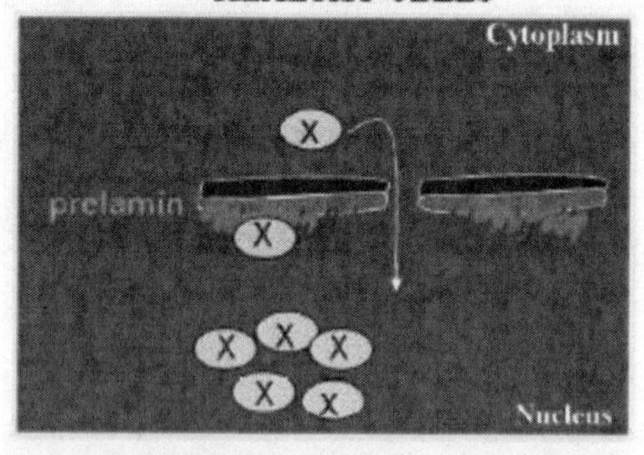

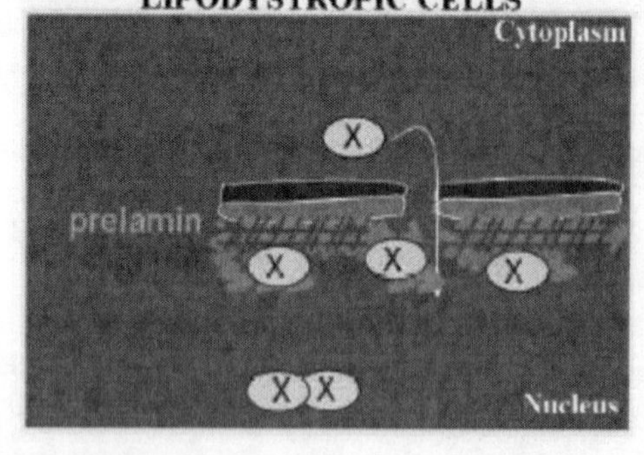

Figure 5. Schematic diagram showing the prelamin A presence in healthy cells and the sequestration of transcription factors by prelamin A accumulation in lipodystrophyc hMSCS.

This interaction between prelamin A and SREBP1 interfered with SREBP1 transcription factor activity, since SREBP1 target genes were dysregulated and adipogenesis was compromised.

Owing to this experimental model based on hMSCs, Sp1 has been found to be essential for hMSCs adipogenesis. This fact has been demonstrated with the utilization of Sp1 specific inhibitor WP631, a bisintercalating anthracycline drug which specifically inhibits Sp1-dependent transcription (Bernot et al., 2010; Mansilla et al., 2004; Mansilla and Portugal, 2008). The inhibition of Sp1 transcription factor activity during hMSCs differentiation to adipocytes resulted in a reduction in their lipid vesicle size, indicating a role for Sp1 in the maintenance and integrity of the lipid vesicles. Thus, Sp1 emerged as a new player in the adipogenesis of hMSCs.

In conclusion, this lipodystrophy model based on hMSCs reported a new mechanism of impaired adipogenesis as a consequence of the interaction of the transcription factor Sp1 and accumulated prelamin A. It should be pointed out that since this human model is based on primary hMSCs offers a new experimental tool to elucidate the pathological mechanisms of *LMNA*-linked lipodystrophies in a biological context that could be very similar to the *in vivo* situation. Finally, this model also provides an ideal system to identify potential targets to generate new therapies and to carry out drug discovery screening for possible lipodystrophy treatments.

Adipogenesis Models Based on Induced Pluripotent Stem Cells (iPSCs)

Although hMSCs-based cell models have proven useful, there are some limitations associated with these cells: they have limited proliferative capacity and variable differentiation potential due to the biological diversity inherent to all individuals (Ryden et al., 2003; Leskela et al., 2006). Regarding *LMNA*-linked lipodystrophies, our hMSCs-based cell model offers an excellent tool to study acquired lipodystrophies. However, if the aim is to study genetic lipodystrophies, iPSCs (induced pluripotent stem cells) based cellular models could be more appropriate. Access to patient skin biopsies to generate iPSCs is easier and less invasive than obtaining hMSCs from bone marrow or adipose tissue from patients.

In 2006, Yamanaka and co-workers first reported the generation of induced pluripotent stem cells (iPSCs) from mouse somatic fibroblasts by the retroviral transduction of four transcription factors: Oct4, Sox2, Klf4 and c-

Myc (Takahashi et al., 2007). It is worth mentioning that due to these studies Dr. Yamanaka was awarded the Nobel Prize for Medicine in 2012, together with Dr. Gurdon.

Human iPSCs have been generated from embryonic, neonatal and adult fibroblasts (Takahashi et al., 2007; Yu et al., 2007; Park et al., 2008b). The concept of utilizing iPSCs to model a human disease in a culture dish is based on the unique capacity of these cells to continuously self-renew (unlimited proliferation ability) and their potential to give rise to at least theoretically, all cell types in the human body. The greatest advantage of iPSC technology is that it allows for the generation of pluripotent cells from any individual in the context of his or her own particular genetic identity, including individuals with sporadic forms of disease and those affected by complex multifactorial diseases of unknown genetic identity. However, the utility of iPSCs is strongly dependent on efficient differentiation protocols that convert cells into relevant adult cell types.

For adipocytes, an efficient differentiation method of inducing iPSCs to form white and brown adipocytes by the inducible expression of specific transcription factors has recently been reported (Ahfeldt T et al., 2012). The iPSCs were first differentiated to MSCs and then transduced with different combinations of transcription factors known to be involved in white or brown adipogenesis. PPARγ alone was able to program cells toward white adipocytes, whereas the combination of PPARγ with C/EBPβ alone or with C/EBPβ and PRDM16 induced differentiation of MSCs into brown adipocytes. Notably, a recent paper has described a method of differentiation of iPSCs to functional classical brown adipocytes using a specific hemopoitein complex, without the need of exogenous gene transfer (Nishio et al., 2012). Both brown and white adipocytes derived from both methods of differentiation exhibited normal adipose tissue function.

The contribution of iPSCs to *LMNA*-linked lipodystrophies was recently demonstrated by two different groups which reported the generation of progeria iPSCs from fibroblasts of progeria patients (Zhang et al., 2011; Liu et al., 2011). They propose this iPSC-based model to study the pathogenesis of both human premature and normal physiological aging. Zhang and co-workers demonstrated that progeria-MSCs derived from iPSCs are more susceptible to stress conditions in general due to progerin accumulation. Liu and co-workers reported a DNA protein kinase (DNAPKcs) as a novel marker of aging. They showed an unknown interaction between progerin and DNAPKcs which is involved in aging related events. The expression of DNAPKcs was downregulated in progeria fibroblasts but restored in HGPS-iPSCs.

Additionally, because progeria patients show lipodystrophy, these iPSC-based models will be of enormous interest to study the pathophysiological mechanisms underlying progerin accumulation in the development of lipodystrophy.

Ultimately, disease models based on iPSCs allow the study of the molecular pathological mechanisms of a given disease in a patient-specific and cell specific context, such as lipodystrophy or progeria. Importantly, these human models provide a valuable tool to identify potential molecular targets for drug screening and discovery with the aim of applying them in clinical trials for treatment of diseases such as *LMNA*-linked lipodystrophies.

Conclusion

Along this chapter it has been demonstrated the relevance of murine (*in vitro* and *in vivo*) models in providing valuable information about the pathophysiology lipodystrohies, in addition to the development of strategies to test potential therapeutics approaches. However, the interspecies differences between mouse and humans associated with adipose tissue biology and the unique features of human stem cells define human experimental models based on these cells as essential tools in the study of lipodystrophies and in the identification of disease markers as well as potential targets for therapeutic strategies. These fact are corroborated by the recent generation of *LMNA*-linked lipodystrophies human models based on stem cells in the last couple of years and by the information provided by these models (Zhang et al., 2011; Liu GH et al., 2011; Ruiz de Eguino et al., 2012).

The relevance of the valuable information provided by these different disease models is clearly noticeable as demonstrated first hand in a recent and first clinical trial for pediatric progeria patients (Gordon et al., 2012) based on a farnesyl transferase inhibitor (FTI). The use of FTIs in this clinical trial was supported by previous evidence obtained from basic research that suggested that blocking prelamin or progerin farnesylation was beneficial for cells affected by progeria. Treatment of progeria mice with FTIs improved some disease phenotypes, such as bone mineralization and weight curves (Yang et al., 2006; Fong et al., 2006). In addition, the FTI treatment decreased the nuclear shape abnormalities in *in vitro* experiments carried out in human progeria fibroblasts (Toth et al., 2005). Importantly, this trial demonstrates the potential of translating basic biomedical and genetics research into human treatment. Thus, the results obtained from years of basic research into the biology and posttranslational processing of lamin A have been indispensable to first generating and finally demonstrating by a clinical trial the hypothesis that farnesylation inhibition might be useful in the treatment of progeria. It should be emphasized that in less than 10 years the molecular origin of progeria unknown until then, due to de novo mutation in *LMNA* gene (Erikson 2003), was discovered and that the completion of the first clinical trial with progeria patients was achieved (Gordon et al., 2012), demonstrating the remarkable importance and the outstanding potential of the basic and translational research approach.

Finally, it is worth to mentioning that during the normal process of aging, fat redistribution is observed with peripheral fat loss and central fat gain (Caron-Devarle et al., 2010).

In fact, this age-related fat redistribution could also be considered a minor form of lipodystrophy (Vigouroux et al., 2011). As has been mentioned in this chapter, *LMNA*-linked lipodystrophies have been clearly associated with premature aging phenotpes as well as the already characterized progeroid syndromes. Since both prelamin A and progerin accumulation have been reported to accumulate in elderly human cells (Ragnauth et al., 2010; Scaffidi and Misteli, 2006b; McClintock et al., 2007), it will be of great interest to analyze the contribution of these accumulated proteins in the normal physiological aging process.

Old age is the major risk factor for many diseases including cancer and cardiovascular and neurodegenerative diseases. As the number of people aged 60 years and older is growing rapidly worldwide, keeping the elderly healthy is a priority for health systems. Aging is an inevitable consequence of life, a natural process with gradual decline of many normal biological functions of cells. However, the molecular mechanisms that cause physiological aging are still not completely understood, most likely because of the complex nature of the aging process.

Thus, human cell models based on stem cells which reproduce the conditions of aging will emerge as an excellent tool to unravel the molecular mechanisms of aging. The elucidation of these mechanisms will contribute to the actual knowledge of human aging processes with the purpose of improving not only the length of life, but especially, the quality of life of those who arrive at old age by increasing the number of disease-free years.

References

Agarwal, A. K., Fryns, J. P., Auchus, R. J., Garg, A. Zinc metalloproteinase, ZMPSTE24, is mutated in mandibuloacral dysplasia. *Hum. Mol. Genet.* 2003;12(16):1995-2001.

Agarwal, A. K., Garg, A. Genetic disorders of adipose tissue development, differentiation, and death. *Annu. Rev. Genomics Hum. Genet.* 2006;7:175-99.

Ahfeldt, T., Schinzel, R. T., Lee, Y. K., Hendrickson, D., Kaplan, A., Lum, D. H., et al., Programming human pluripotent stem cells into white and brown adipocytes. *Nat. Cell Biol.* 2012;14(2):209-19.

Ahima, R. S., Flier, J. S. Adipose tissue as an endocrine organ. *Trends Endocrinol. Metab.* 2000;11(8):327-32.

Ailhaud, G., Grimaldi, P., Négrel, R. A molecular view of adipose tissue. *Int. J. Obes. Relat. Metab. Disord.* 1992;16 Suppl. 2:S17-21.

Araújo-Vilar, D., Lattanzi, G., González-Méndez, B., Costa-Freitas, A. T., Prieto, D., Columbaro, M., et al., Site-dependent differences in both prelamin A and adipogenic genes in subcutaneous adipose tissue of patients with type 2 familial partial lipodystrophy. *J. Med. Genet.* 2009;46 (1):40-8.

Araújo-Vilar, D., Victoria, B., González-Méndez, B., Barreiro, F., Fernández-Rodríguez, B., Cereijo, R., et al., Histological and molecular features of lipomatous and nonlipomatous adipose tissue in familial partial lipodystrophy caused by LMNA mutations. *Clin. Endocrinol.* (Oxf.). 2012;76(6):816-24.

Barbatelli, G., Murano, I., Madsen, L., Hao, Q., Jimenez, M., Kristiansen, K., et al., The emergence of cold-induced brown adipocytes in mouse white fat depots is determined predominantly by white to brown adipocyte transdifferentiation. *Am. J. Physiol. Endocrinol. Metab.* 2010;298(6):E12 44-53.

Bartelt, A., Bruns, O. T., Reimer, R., Hohenberg, H., Ittrich, H., Peldschus, K., et al., Brown adipose tissue activity controls triglyceride clearance. *Nat. Med.* 2011;17(2):200-5.

Bastard, J. P., Caron, M., Vidal, H., Jan, V., Auclair, M., Vigouroux, C., et al., Association between altered expression of adipogenic factor SREBP1 in lipoatrophic adipose tissue from HIV-1-infected patients and abnormal adipocyte differentiation and insulin resistance. *Lancet.* 2002;359 (9311): 1026-31.

Beck, L. A., Hosick, T. J., Sinensky, M. Isoprenylation is required for the processing of the lamin A precursor. *J. Cell Biol.* 1990;110(5):1489-99.

Benavides, F., Guénet, J. L. Murine *Medicina* (B Aires). 2001;61(2):215-31.

Bergo, M. O., Gavino, B., Ross, J., Schmidt, W. K., Hong, C., Kendall, L. V., et al., Zmpste24 deficiency in mice causes spontaneous bone fractures, muscle weakness, and a prelamin A processing defect. *Proc. Natl. Acad. Sci. US* 2002;99(20):13049-54.

Bernot, D., Barruet, E., Poggi, M., Bonardo, B., Alessi, M. C., Peiretti, F. Down-regulation of tissue inhibitor of metalloproteinase-3 (TIMP-3) expression is necessary for adipocyte differentiation. *J. Biol. Chem.* 2010; 285(9):6508-14.

Biamonti, G., Giacca, M., Perini, G., Contreas, G., Zentilin, L., Weighardt, F., et al., The gene for a novel human lamin maps at a highly transcribed locus of chromosome 19 which replicates at the onset of S-phase. *Mol. Cell Biol.* 1992;12(8):3499-506.

Bidault, G., Vatier, C., Capeau, J., Vigouroux, C., Béréziat, V. LMNA-linked lipodystrophies: from altered fat distribution to cellular alterations. *Biochem. Soc. Trans.* 2011;39(6):1752-7.

Bione, S., Maestrini, E., Rivella, S., Mancini, M., Regis, S., Romeo, G., et al., Identification of a novel X-linked gene responsible for Emery-Dreifuss muscular dystrophy. *Nat. Genet.* 1994;8(4):323-7.

Bonne, G., Di Barletta, M. R., Varnous, S., Bécane, H. M., Hammouda, E. H., Merlini, L., et al., Mutations in the gene encoding lamin A/C cause autosomal dominant Emery-Dreifuss muscular dystrophy. *Nat. Genet.* 1999;21(3):285-8.

Bremer, A. A., Devaraj, S., Afify, A., Jialal, I. Adipose tissue dysregulation in patients with metabolic syndrome. *J. Clin. Endocrinol. Metab.* 2011;96 (11):E1782-8.

Broers, J. L., Ramaekers, F. C., Bonne, G., Yaou, R. B., Hutchison, C. J. Nuclear lamins: laminopathies and their role in premature ageing. *Physiol. Rev.* 2006;86(3):967-1008.

Cannon, B., Nedergaard, J. Brown adipose tissue: function and physiological significance. *Physiol. Rev.* 2004;84(1):277-359.

Cao, H., Hegele, R. A. Nuclear lamin A/C R482Q mutation in canadian kindreds with Dunnigan-type familial partial lipodystrophy. *Hum. Mol. Genet.* 2000;9(1):109-12.

Cao, K., Capell, B. C., Erdos, M. R., Djabali, K., Collins, F. S. A lamin A protein isoform overexpressed in Hutchinson-Gilford progeria syndrome interferes with mitosis in progeria and normal cells. *Proc. Natl. Acad. Sci. US* 2007;104(12):4949-54.

Cao, K., Blair, C. D., Faddah, D. A., Kieckhaefer, J. E., Olive, M., Erdos, M. R., Nabel, E. G., Collins, F.S. Progerin and telomere. *J. Clin. Invest.* 2011 July 1; 121(7): 2833–2844.

Capanni, C., Mattioli, E., Columbaro, M., Lucarelli, E., Parnaik, V. K., Novelli, G., et al., Altered pre-lamin A processing is a common mechanism leading to lipodystrophy. *Hum. Mol. Genet.* 2005;14(11): 1489-502.

Capanni, C., Squarzoni, S., Cenni, V., D'Apice, M. R., Gambineri, A., Novelli, G., et al., Familial partial lipodystrophy, mandibuloacral dysplasia and restrictive dermopathy feature barrier-to-autointegration factor (BAF) nuclear redistribution. *Cell Cycle.* 2012;11(19):3568-77.

Capeau, J., Magré, J., Caron-Debarle, M., Lagathu, C., Antoine, B., Béréziat, V., et al., Human lipodystrophies: genetic and acquired diseases of adipose tissue. *Endocr. Dev.* 2010;19:1-20.

Capell, B. C., Collins, F. S. Human laminopathies: nuclei gone genetically awry. *Nat. Rev. Genet.* 2006;7(12):940-52.

Caron, M., Auclair, M., Vigouroux, C., Glorian, M., Forest, C., Capeau, J. The HIV protease inhibitor indinavir impairs sterol regulatory element-binding protein-1 intranuclear localization, inhibits preadipocyte differentiation, and induces insulin resistance. *Diabetes.* 2001;50(6):1378-88.

Caron, M., Auclair, M., Sterlingot, H., Kornprobst, M., Capeau, J. Some HIV protease inhibitors alter lamin A/C maturation and stability, SREBP-1 nuclear localization and adipocyte differentiation. *AIDS* 2003 Nov. 21;17 (17):2437-44.

Caron, M., Auclair, M., Donadille, B., Béréziat, V., Guerci, B., Laville, M., et al., Human lipodystrophies linked to mutations in A-type lamins and to HIV protease inhibitor therapy are both associated with prelamin A accumulation, oxidative stress and premature cellular senescence. *Cell Death Differ.* 2007a;14(10):1759-67.

Caron, M., Vigouroux, C., Bastard, J. P., Capeau, J. Adipocyte dysfunction in response to antiretroviral therapy: clinical, tissue and in-vitro studies. *Curr. Opin. HIV AIDS.* 2007b;2(4):268-73.

Caron, M., Auclairt, M., Vissian, A., Vigouroux, C., Capeau, J. Contribution of mitochondrial dysfunction and oxidative stress *Antivir. Ther.* 2008;13(1):27-38.

Caron-Debarle, M., Lagathu, C., Boccara, F., Vigouroux, C., Capeau, J. HIV-associated lipodystrophy: from fat injury to premature aging. *Trends Mol. Med.* 2010;16(5):218-29.

Chartoumpekis, D. V., Habeos, I. G., Ziros, P. G., Psyrogiannis, A. I., Kyriazopoulou, V. E., Papavassiliou, A. G. Brown adipose tissue responds to cold and adrenergic stimulation by induction of FGF21. *Mol. Med.* 2011;17(7-8):736-40.

Cinti, S. The adipose organ. *Prostaglandins Leukot. Essent. Fatty Acids.* 2005; 73(1):9-15.

Coffinier, C., Hudon, S. E., Farber, E. A., Chang, S. Y., Hrycyna, C. A., Young, S. G., et al., HIV protease inhibitors block the zinc metalloproteinase ZMPSTE24 and lead to an accumulation of prelamin A in cells. *Proc. Natl. Acad. Sci. US* 2007;104(33):13432-7.

Coffinier, C., Jung, H. J., Li, Z., Nobumori, C., Yun, U. J., Farber, E. A., et al., Direct synthesis of lamin A, bypassing prelamin a processing, causes misshapen nuclei in fibroblasts but no detectable pathology in mice. *J. Biol. Chem.* 2010;285(27):20818-26.

Constantinescu, D., Gray, H. L., Sammak, P. J., Schatten, G. P., Csoka, A. B. Lamin A/C expression is a marker of mouse and human embryonic stem cell differentiation. *Stem Cells.* 2006;24(1):177-85.

Crowe, S. M., Westhorpe, C. L., Mukhamedova, N., Jaworowski, A., Sviridov, D., Bukrinsky, M. The macrophage: the intersection between HIV *J. Leukoc. Biol.* 2010 Apr.;87(4):589-98.

Cutler, D. A., Sullivan, T., Marcus-Samuels, B., Stewart, C. L., Reitman, M. L. Characterization of adiposity and metabolism in Lmna-deficient mice. *Biochem. Biophys. Res. Commun.* 2002;291(3):522-7.

Cypess, A. M., Lehman, S., Williams, G., Tal, I., Rodman, D., Goldfine, A. B., et al., Identification and importance of brown adipose tissue in adult humans. *N Engl. J. Med.* 2009;360(15):1509-17.

Dai, Q., Choy, E., Chiu, V., Romano, J., Slivka, S. R., Steitz, S. A., et al., Mammalian prenylcysteine carboxyl methyltransferase is in the endoplasmic reticulum. *J. Biol. Chem.* 1998;273(24):15030-4.

Danforth, E. Failure of adipocyte differentiation causes type II diabetes mellitus? *Nat. Genet.* 2000;26(1):13.

Darimont, C., Macé, K. Immortalization of human preadipocytes. *Biochimie.* 2003;85(12):1231-3.

De Sandre-Giovannoli, A., Bernard, R., Cau, P., Navarro, C., Amiel, J., Boccaccio, I., et al., Lamin a truncation in Hutchinson-Gilford progeria. *Science.* 2003;300(5628):2055.

Dechat, T., Pfleghaar, K., Sengupta, K., Shimi, T., Shumaker, D. K., Solimando, L., et al., Nuclear lamins: major factors in the structural organization and function of the nucleus and chromatin. *Genes Dev.* 2008; 22(7):832-53.

Dowell, P., Flexner, C., Kwiterovich, P. O., Lane, M. D. Suppression of preadipocyte differentiation and promotion of adipocyte death by HIV protease inhibitors. *J. Biol. Chem.* 2000;275(52):41325-32.

Dunnigan, M. G., Cochrane, M. A., Kelly, A., Scott, J. W. Familial lipoatrophic diabetes with dominant transmission. A new syndrome. *Q. J. Med.* 1974;43(169):33-48.

Elabd, C., Chiellini, C., Carmona, M., Galitzky, J., Cochet, O., Petersen, R., et al., Human multipotent adipose-derived stem cells differentiate into functional brown adipocytes. *Stem Cells.* 2009;27(11):2753-60.

Eriksson, M., Brown, W. T., Gordon, L. B., Glynn, M. W., Singer, J., Scott, L., Erdos, M. R., Robbins, C. M., Moses, T. Y., Berglund, P., et al., Recurrent de novo point mutations in lamin A cause Hutchinson-Gilford progeria syndrome. *Nature.* 2003 May 15;423(6937):293-8.

Espada, J., Varela, I., Flores, I., Ugalde, A. P., Cadiñanos, J., Pendás, A. M., et al., Nuclear envelope defects cause stem cell dysfunction in premature-aging mice. *J. Cell Biol.* 2008;181(1):27-35.

Fatkin, D., MacRae, C., Sasaki, T., Wolff, M. R., Porcu, M., Frenneaux, M., et al., Missense mutations in the rod domain of the lamin A/C gene as causes of dilated cardiomyopathy and conduction-system disease. *N Engl. J. Med.* 1999;341(23):1715-24.

Fève, B. Adipogenesis: cellular and molecular aspects. *Best Pract. Res. Clin. Endocrinol. Metab.* 2005;19(4):483-99.

Filesi, I., Gullotta, F., Lattanzi, G., D'Apice, M. R., Capanni, C., Nardone, A. M., et al., Alterations of nuclear envelope and chromatin organization in mandibuloacral dysplasia, a rare form of laminopathy. *Physiol. Genomics.* 2005;23(2):150-8.

Fong, L. G., Frost, D., Meta, M., Qiao, X., Yang, S. H., Coffinier, C., et al., A protein farnesyltransferase inhibitor ameliorates disease in a mouse model of progeria. *Science.* 2006;311(5767):1621-3.

Fong, L. G., Ng, J. K., Meta, M., Coté, N., Yang, S. H., Stewart, C. L., et al., Heterozygosity for Lmna deficiency eliminates the progeria-like phenotypes in Zmpste24-deficient mice. *Proc. Natl. Acad. Sci. US* 2004; 101(52):18111-6.

Fontaine, C., Cousin, W., Plaisant, M., Dani, C., Peraldi, P. Hedgehog signaling alters adipocyte maturation of human mesenchymal stem cells. *Stem Cells*. 2008;26(4):1037-46.

Frayn, K. N., Karpe, F., Fielding, B. A., Macdonald, I. A., Coppack, S. W. Integrative physiology of human adipose tissue. *Int. J. Obes. Relat. Metab. Disord.* 2003;27(8):875-88.

Frühbeck, G., Gómez-Ambrosi, J., Muruzábal, F. J., Burrell, M. A. The adipocyte: a model for integration of endocrine and metabolic signaling in energy metabolism regulation. *Am. J. Physiol. Endocrinol. Metab.* 2001; 280(6):E827-47.

Furukawa, K., Inagaki, H., Hotta, Y. Identification and cloning of an mRNA coding for a germ cell-specific A-type lamin in mice. *Exp. Cell Res.* 1994; 212(2):426-30.

Gallant, J. E., Pham, P. A. Tenofovir disoproxil fumarate (Viread) for the treatment of HIV infection. *Expert Rev. Anti Infect. Ther.* 2003 Oct.;1(3): 415-22.

Garg, A. Clinical review#: Lipodystrophies: genetic and acquired body fat disorders. *J. Clin. Endocrinol. Metab*. 2011;96(11):3313-25.

Gavrilova, O., Marcus-Samuels, B., Graham, D., Kim, J. K., Shulman, G. I., Castle, A. L., et al., Surgical implantation of adipose tissue reverses diabetes in lipoatrophic mice. *J. Clin. Invest*. 2000;105(3):271-8.

Gesta, S., Tseng, Y. H., Kahn, C. R. Developmental origin of fat: tracking obesity to its source. *Cell*. 2007;131(2):242-56.

Goldman, R. D., Shumaker, D. K., Erdos, M. R., Eriksson, M., Goldman, A. E., Gordon, L. B., et al., Accumulation of mutant lamin A causes progressive changes in nuclear architecture in Hutchinson-Gilford progeria syndrome. *Proc. Natl. Acad. Sci. US* 2004;101(24):8963-8.

González, J. M., Andrés, V. Synthesis, transport and incorporation into the nuclear envelope of A-type lamins and inner nuclear membrane proteins. *Biochem. Soc. Trans.* 2011;39(6):1758-63.

González, J. M., Navarro-Puche, A., Casar, B., Crespo, P., Andrés, V. Fast regulation of AP-1 activity through interaction of lamin A/C, ERK1/2, and c-Fos at the nuclear envelope. *J. Cell Biol*. 2008;183(4):653-66.

Gordon, L. B., Kleinman, M. E., Miller, D. T., Neuberg, D. S., Giobbie-Hurder, A., Gerhard-Herman, M., et al., Clinical trial of a farnesyltransferase inhibitor in children with Hutchinson-Gilford progeria syndrome. *Proc. Natl. Acad. Sci. US* 2012;109(41):16666-71.

Grinspoon, S., Carr, A. Cardiovascular risk and body-fat abnormalities in HIV-infected adults. *N Engl. J. Med*. 2005 Jan. 6;352(1):48-62.

Hammond, E., Nolan, D. Adipose tissue inflammation and altered adipokine and cytokine production in antiretroviral therapy-associated lipodystrophy. *Curr. Opin. HIV AIDS*. 2007;2(4):274-81.

Hansen, J. B., Kristiansen, K. Pocket proteins control white versus brown fat cell differentiation. *Cell Cycle*. 2006;5(4):341-2.

Hegele, R. A. Molecular basis of partial lipodystrophy and prospects for therapy. *Trends Mol. Med*. 2001;7(3):121-6.

Helbling-Leclerc, A., Bonne, G., Schwartz, K. Emery-Dreifuss muscular dystrophy. *Eur. J. Hum. Genet*. 2002;10(3):157-61.

Hernandez, L., Roux, K. J., Wong, E. S., Mounkes, L. C., Mutalif, R., Navasankari, R., et al., Functional coupling between the extracellular matrix and nuclear lamina by Wnt signaling in progeria. *Dev. Cell*. 2010; 19(3):413-25.

Himms-Hagen, J. Physiological roles of the leptin endocrine system: differences between mice and humans. *Crit. Rev. Clin. Lab. Sci*. 1999;36 (6):575-655.

Hondares, E., Rosell, M., Gonzalez, F. J., Giralt, M., Iglesias, R., Villarroya, F. Hepatic FGF21 expression is induced at birth via PPARalpha in response to milk intake and contributes to thermogenic activation of neonatal brown fat. *Cell Metab*. 2010;11(3):206-12.

Hugo, E. R., Brandebourg, T. D., Comstock, C. E., Gersin, K. S., Sussman, J. J., Ben-Jonathan, N. LS14: a novel human adipocyte cell line that produces prolactin. *Endocrinology*. 2006;147(1):306-13.

Imai, T., Takakuwa, R., Marchand, S., Dentz, E., Bornert, J. M., Messaddeq, N., et al., Peroxisome proliferator-activated receptor gamma is required in mature white and brown adipocytes for their survival in the mouse. *Proc. Natl. Acad. Sci. US* 2004;101(13):4543-7.

Ivorra, C., Kubicek, M., González, J. M., Sanz-González, S. M., Alvarez-Barrientos, A., O'Connor, J. E., et al., A mechanism of AP-1 suppression through interaction of c-Fos with lamin A/C. *Genes Dev*. 2006;20(3):307-20.

Jacobson, J. M. Immune-based therapies: an adjunct to antiretroviral treatment. *Curr. HIV* 2005 Jun.;2(2):90-7.

Jain, R. G., Lenhard, J. M. Select HIV protease inhibitors alter bone and fat metabolism ex vivo. *J. Biol. Chem*. 2002;277(22):19247-50.

Jeninga, E. H., Kalkhoven, E. Central players in inherited lipodystrophies. *Trends Endocrinol. Metab*. 2010;21(10):581-8.

Kazantzis, M., Takahashi, V., Hinkle, J., Kota, S., Zilberfarb, V., Issad, T., et al., PAZ6 cells constitute a representative model for human brown pre-adipocytes. *Front Endocrinol.* (Lausanne). 2012;3:13.

Kershaw, E. E., Flier, J. S. Adipose tissue as an endocrine organ. *J. Clin. Endocrinol. Metab.* 2004;89(6):2548-56.

Klingenspor, M. Cold-induced recruitment of brown adipose tissue thermogenesis. *Exp. Physiol.* 2003;88(1):141-8.

Köbberling, J., Dunnigan, M. G. Familial partial lipodystrophy: two types of an X linked dominant syndrome, lethal in the hemizygous state. *J. Med. Genet.* 1986;23(2):120-7.

Kratz, M., Purnell, J. Q., Breen, P. A., Thomas, K. K., Utzschneider, K. M., Carr, D. B., et al., Reduced adipogenic gene expression in thigh adipose tissue precedes human immunodeficiency virus-associated lipoatrophy. *J. Clin. Endocrinol. Metab.* 2008;93(3):959-66.

Kubben, N., Voncken, J. W., Konings, G., van Weeghel, M., van den Hoogenhof, M. M., Gijbels, M., et al., Post-natal myogenic and adipogenic developmental: defects and metabolic impairment upon loss of A-type lamins. *Nucleus.* 2011;2(3):195-207.

Lafontan, M., Berlan, M. Fat cell adrenergic receptors and the control of white and brown fat cell function. *J. Lipid Res.* 1993;34(7):1057-91.

Lammerding, J., Schulze, P. C., Takahashi, T., Kozlov, S., Sullivan, T., Kamm, R. D., et al., Lamin A/C deficiency causes defective nuclear mechanics and mechanotransduction. *J. Clin. Invest.* 2004;113(3):370-8.

Lanktree, M., Cao, H., Rabkin, S. W., Hanna, A., Hegele, R. A. Novel LMNA mutations seen in patients with familial partial lipodystrophy subtype 2 (FPLD2; MIM 151660). *Clin. Genet.* 2007 Feb;71(2):183-6.

Lee, D. C., Welton, K. L., Smith, E. D., Kennedy, B. K. A-type nuclear lamins act as transcriptional repressors when targeted to promoters. *Exp. Cell Res.* 2009;315(6):996-1007.

Lenhard, J. M., Furfine, E. S., Jain, R. G., Ittoop, O., Orband-Miller, L. A., Blanchard, S. G., et al., HIV protease inhibitors block adipogenesis and increase lipolysis in vitro. *Antiviral Res.* 2000;47(2):121-9.

Leskelä, H. V., Olkku, A., Lehtonen, S., Mahonen, A., Koivunen, J., Turpeinen, M., et al., Estrogen receptor alpha genotype confers interindividual variability of response to estrogen and testosterone in mesenchymal-stem-cell-derived osteoblasts. *Bone.* 2006;39(5):1026-34.

Lichtenstein, K. A. Redefining lipodystrophy syndrome: risks and impact on clinical decision making. *J. Acquir. Immune Defic. Syndr.* 2005;39(4): 395-400.

Lin, F., Worman, H. J. Structural organization of the human gene encoding nuclear lamin A and nuclear lamin C. *J. Biol. Chem.* 1993;268(22):16321-6.

Lin, F., Worman, H. J. Structural organization of the human gene (LMNB1) encoding nuclear lamin B1. *Genomics*. 1995;27(2):230-6.

Liu, G. H., Barkho, B. Z., Ruiz, S., Diep, D., Qu, J., Yang, S. L,. et al., Recapitulation of premature ageing with iPSCs from Hutchinson-Gilford progeria syndrome. *Nature*. 2011;472(7342):221-5.

Mackay, D. L., Tesar, P. J., Liang, L. N., Haynesworth, S. E. Characterizing medullary and human mesenchymal stem cell-derived adipocytes. *J. Cell Physiol.* 2006;207(3):722-8.

Maher, B., Alfirevic, A., Vilar, F. J., Wilkins, E. G., Park, B. K., Pirmohamed, M. TNF *AIDS*. 2002 Oct. 18;16(15):2013-8.

Mansilla, E., Díaz Aquino, V., Zambón, D., Marin, G. H., Mártire, K., Roque, G., et al., Could metabolic syndrome, lipodystrophy, and aging be mesenchymal stem cell exhaustion syndromes? *Stem Cells Int*. 2011;2011: 943216.

Mansilla, S., Portugal, J. Sp1 transcription factor as a target for anthracyclines: effects on gene transcription. *Biochimie*. 2008;90(7):976-87.

Mansilla, S., Priebe, W., Portugal, J. Sp1-targeted inhibition of gene transcription by WP631 in transfected lymphocytes. *Biochemistry*. 2004; 43(23):7584-92.

Mariman, E. C., Wang, P. Adipocyte extracellular matrix composition, dynamics and role in obesity. *Cell Mol. Life Sci*. 2010;67(8):1277-92.

Maske, C. P., Hollinshead, M. S., Higbee, N. C., Bergo, M. O., Young, S. G., Vaux, D. J. A carboxyl-terminal interaction of lamin B1 is dependent on the CAAX endoprotease Rce1 and carboxymethylation. *J. Cell Biol.* 2003; 162(7):1223-32.

Mattsson, C. L., Csikasz, R. I., Chernogubova, E., Yamamoto, D. L., Hogberg, H. T., Amri, E. Z., et al., β1-Adrenergic receptors increase UCP1 in human MADS brown adipocytes and rescue cold-acclimated β3-adrenergic receptor-knockout mice via nonshivering thermogenesis. *Am. J. Physiol. Endocrinol. Metab.* 2011;301(6):E1108-18.

Maurin, T., Saillan-Barreau, C., Cousin, B., Casteilla, L., Doglio, A., Pénicaud, L. Tumor necrosis *Exp. Cell Res.* 2005 Apr. 1;304(2): 544-51.

McClintock, D., Ratner, D., Lokuge, M., Owens, D. M., Gordon, L. B., Collins, F. S., et al., The mutant form of lamin A that causes Hutchinson-Gilford progeria is a biomarker of cellular aging in human skin. *PLoS One*. 2007;2(12):e1269.

Minguell, J. J., Erices, A., Conget, P. Mesenchymal stem cells. *Exp. Biol. Med.* (Maywood). 2001;226(6):507-20.

Moitra, J., Mason, M. M., Olive, M., Krylov, D., Gavrilova, O., Marcus-Samuels, B., et al., Life without white fat: a transgenic mouse. *Genes Dev.* 1998;12(20):3168-81.

Monajemi, H., Stroes, E., Hegele, R. A., Fliers, E. Inherited lipodystrophies and the metabolic syndrome. *Clin. Endocrinol.* (Oxf.). 2007;67(4):479-84.

Morganstein, D. L., Wu, P., Mane, M. R., Fisk, N. M., White, R., Parker, M. G. Human fetal mesenchymal stem cells differentiate into brown and white adipocytes: a role for ERRalpha in human UCP1 expression. *Cell Res.* 2010;20(4):434-44.

Mory, P. B., Crispim, F., Freire, M. B., Salles, J. E., Valério, C. M., Godoy-Matos, A. F., et al., Phenotypic diversity in patients with lipodystrophy associated with LMNA mutations. *Eur. J. Endocrinol.* 2012;167(3):423-31.

Mounkes, L. C., Kozlov, S., Hernandez, L., Sullivan, T., Stewart, C. L. A progeroid syndrome in mice is caused by defects in A-type lamins. *Nature*. 2003;423(6937):298-301.

Moyle, G. Mechanisms of HIV *Antivir. Ther.* 2005;10 Suppl. 2:M47-52. Review.

Murphy, M. P. Investigating mitochondrial radical production using targeted probes. *Biochem. Soc. Trans.* 2004 Dec;32(Pt 6):1011-4.

Nakajima, N., Abe, K. Genomic structure of the mouse A-type lamin gene locus encoding somatic and germ cell-specific lamins. *FEBS Lett.* 1995; 365(2-3):108-14.

Navarro, C. L., Cadiñanos, J., De Sandre-Giovannoli, A., Bernard, R., Courrier, S., Boccaccio, I., et al., Loss of ZMPSTE24 (FACE-1) causes autosomal recessive restrictive dermopathy and accumulation of Lamin A precursors. *Hum. Mol. Genet.* 2005;14(11):1503-13.

Nedergaard, J., Bengtsson, T., Cannon, B. Unexpected evidence for active brown adipose tissue in adult humans. *Am. J. Physiol. Endocrinol. Metab.* 2007;293(2):E444-52.

Nikolova, V., Leimena, C., McMahon, A. C., Tan, J. C., Chandar, S., Jogia, D., et al., Defects in nuclear structure and function promote dilated cardiomyopathy in lamin A/C-deficient mice. *J. Clin. Invest.* 2004;113 (3):357-69.

Nishio, M., Yoneshiro, T., Nakahara, M., Suzuki, S., Saeki, K., Hasegawa, M., et al., Production of functional classical brown adipocytes from human pluripotent stem cells using specific hemopoietin cocktail without gene transfer. *Cell Metab.* 2012;16(3):394-406.

Novelli, G., Muchir, A., Sangiuolo, F., Helbling-Leclerc, A., D'Apice, M. R., Massart, C., Capon, F., Sbraccia, P., Federici, M., Lauro, R., Tudisco, C., Pallotta, R., Scarano, G., Dallapiccola, B., Merlini, L., Bonne, G. Mandibuloacral dysplasia *Am. J. Hum. Genet.* 2002 Aug.;71(2):426-31.

Osorio, F. G., Navarro, C. L., Cadiñanos, J., López-Mejía, I. C., Quirós, P. M., Bartoli, C., et al., Splicing-directed therapy in a new mouse model of human accelerated aging. *Sci. Transl. Med.* 2011;3(106):106ra7.

Pacenti, M., Barzon, L., Favaretto, F., Fincati, K., Romano, S., Milan, G., et al., Microarray analysis during adipogenesis identifies new genes altered by antiretroviral drugs. *AIDS.* 2006;20(13):1691-705.

Pajvani, U. B., Trujillo, M. E., Combs, T. P., Iyengar, P., Jelicks, L., Roth, K. A., et al., Fat apoptosis through targeted activation of caspase 8: a new mouse model of inducible and reversible lipoatrophy. *Nat. Med.* 2005;11 (7):797-803.

Palchetti, C. Z., Patin, R. V., Gouvêa, AeF, Szejnfeld, V. L., Succi, R. C., Oliveira, F. L. Body composition and lipodystrophy in prepubertal HIV-infected children. *Braz. J. Infect. Dis.* 2013;17(1):1-6.

Park, I. H., Arora, N., Huo, H., Maherali, N., Ahfeldt, T, Shimamura A, et al., Disease-specific induced pluripotent stem cells. *Cell.* 2008b;134(5):877-86.

Park, I. H., Zhao, R., West, J. A., Yabuuchi, A., Huo, H., Ince, T. A., et al., Reprogramming of human somatic cells to pluripotency with defined factors. *Nature.* 2008;451(7175):141-6.

Park, K. W., Halperin, D. S., Tontonoz, P. Before they were fat: adipocyte progenitors. *Cell Metab.* 2008a;8(6):454-7.

Parra, M. K., Gee, S., Mohandas, N., Conboy, J. G. Efficient in vivo manipulation of alternative pre-mRNA splicing events using antisense morpholinos in mice. *J. Biol. Chem.* 2011;286(8):6033-9.

Pendás, A. M., Zhou, Z., Cadiñanos, J., Freije, J. M., Wang, J., Hultenby, K., et al., Defective prelamin A processing and muscular and adipocyte alterations in Zmpste24 metalloproteinase-deficient mice. *Nat. Genet.* 2002;31(1):94-9.

Petrovic, N., Walden, T. B., Shabalina, I. G., Timmons, J. A., Cannon, B., Nedergaard, J. Chronic peroxisome proliferator-activated receptor gamma (PPARgamma) activation of epididymally derived white adipocyte cultures reveals a population of thermogenically competent, UCP1-

containing adipocytes molecularly distinct from classic brown adipocytes. *J. Biol. Chem.* 2010;285(10):7153-64.

Pisani, D. F., Djedaini, M., Beranger, G. E., Elabd, C., Scheideler, M., Ailhaud, G., et al., Differentiation of Human Adipose-Derived Stem Cells into "Brite" (Brown-in-White) Adipocytes. *Front Endocrinol.* (Lausanne). 2011;2:87.

Pittenger, M. F., Mackay, A. M., Beck, S. C., Jaiswal, R. K., Douglas, R., Mosca, J. D., et al., Multilineage potential of adult human mesenchymal stem cells. *Science.* 1999;284(5411):143-7.

Plasilova, M., Chattopadhyay, C., Pal, P., Schaub, N. A., Buechner, S. A., Mueller, H., Miny, P., Ghosh, A., Heinimann, K. Homozygous missense mutation in the lamin A/C gene causes autosomal recessive Hutchinson-Gilford progeria syndrome. *J. Med. Genet.* 2004 Aug;41(8):609-14.

Ragnauth, C. D., Warren, D. T., Liu, Y., McNair, R., Tajsic, T., Figg, N., et al., Prelamin A acts to accelerate smooth muscle cell senescence and is a novel biomarker of human vascular aging. *Circulation.* 2010;121(20): 2200-10.

Reddy, K. L., Zullo, J. M., Bertolino, E., Singh, H. Transcriptional repression mediated by repositioning of genes to the nuclear lamina. *Nature.* 2008; 452(7184):243-7.

Riordan, N. H., Ichim, T. E., Min, W. P., Wang, H., Solano, F., Lara, F., et al., Non-expanded adipose stromal vascular fraction cell therapy for multiple sclerosis. *J. Transl. Med.* 2009;7:29.

Rodríguez, A. M., Elabd, C., Delteil, F., Astier, J., Vernochet, C., Saint-Marc, P., et al., Adipocyte differentiation of multipotent cells established from human adipose tissue. *Biochem. Biophys. Res. Commun.* 2004;315(2):255-63.

Rosen, E. D., Spiegelman, B. M. Molecular regulation of adipogenesis. *Annu. Rev. Cell Dev. Biol.* 2000;16:145-71.

Rosen, E. D., Spiegelman, B. M. Adipocytes as regulators of energy balance and glucose homeostasis. *Nature.* 2006;444(7121):847-53.

Rosen, E. D., Walkey, C. J., Puigserver, P., Spiegelman, B. M. Transcriptional regulation of adipogenesis. *Genes Dev.* 2000;14(11):1293-307.

Ruiz de Eguino, G., Infante, A., Schlangen, K., Aransay, A. M., Fullaondo, A., Soriano, M., et al., Sp1 transcription factor interaction with accumulated prelamin a impairs adipose lineage differentiation in human mesenchymal stem cells: essential role of sp1 in the integrity of lipid vesicles. *Stem Cells Transl. Med.* 2012;1(4):309-21.

Rydén, M., Dicker, A., Götherström, C., Aström, G., Tammik, C., Arner, P., et al., Functional characterization of human mesenchymal stem cell-derived adipocytes. *Biochem. Biophys. Res. Commun.* 2003;311(2):391-7.

Sanchez-Gurmaches, J., Hung, C. M., Sparks, C. A., Tang, Y., Li, H., Guertin, D. A. PTEN loss in the Myf5 lineage redistributes body fat and reveals subsets of white adipocytes that arise from Myf5 precursors. *Cell Metab.* 2012;16(3):348-62.

Savage, D. B. Mouse models of inherited lipodystrophy. *Dis. Model Mech.* 2009;2(11-12):554-62.

Scaffidi, P., Misteli, T. Good news in the nuclear envelope: loss of lamin A might be a gain. *J. Clin. Invest.* 2006a;116(3):632-4.

Scaffidi, P., Misteli, T. Lamin A-dependent nuclear defects in human aging. *Science.* 2006b;312(5776):1059-63.

Scaffidi, P., Misteli, T. Lamin A-dependent misregulation of adult stem cells associated with accelerated ageing. *Nat. Cell Biol.* 2008;10(4):452-9.

Seale, P., Bjork, B., Yang, W., Kajimura, S., Chin, S., Kuang, S., et al., PRDM16 controls a brown fat/skeletal muscle switch. *Nature.* 2008;454 (7207):961-7.

Seip, M., Trygstad, O. Generalized lipodystrophy, congenital and acquired (lipoatrophy). *Acta Paediatr.* Suppl. 1996;413:2-28.

Shackleton, S., Lloyd, D. J., Jackson, S. N., Evans, R., Niermeijer, M. F., Singh, B. M., et al., LMNA, encoding lamin A/C, is mutated in partial lipodystrophy. *Nat. Genet.* 2000;24(2):153-6.

Shao, W., Espenshade, P. J. Expanding roles for SREBP in metabolism. *Cell Metab.* 2012;16(4):414-9.

Shimomura, I., Hammer, R. E., Richardson, J. A., Ikemoto, S., Bashmakov, Y., Goldstein, J. L., Brown, M. S. Insulin resistance *Genes Dev.* 1998 Oct. 15;12 (20):3182-94.

Sinensky, M., Fantle, K., Trujillo, M., McLain, T., Kupfer, A., Dalton, M. The processing pathway of prelamin A. *J. Cell Sci.* 1994;107 (Pt 1):61-7.

Speckman, R. A., Garg, A., Du, F., Bennett, L., Veile, R., Arioglu, E., Taylor, S. I., Lovett, M., Bowcock, A. M. Mutational and haplotype analyses of families with familial partial lipodystrophy (Dunnigan variety) reveal recurrent missense mutations in the globular C-terminal domain of lamin A/C. A*m. J. Hum. Genet.* 2000 Apr.; 66(4):1192-8.

Spiegelman, B. M., Flier, J. S. Obesity and the regulation of energy balance. *Cell.* 2001;104(4):531-43.

Stewart, C., Burke, B. Teratocarcinoma stem cells and early mouse embryos contain only a single major lamin polypeptide closely resembling lamin B. *Cell.* 1987;51(3):383-92.

Sullivan, T., Escalante-Alcalde, D., Bhatt, H., Anver, M., Bhat, N., Nagashima, K., et al., Loss of A-type lamin expression compromises nuclear envelope integrity leading to muscular dystrophy. *J. Cell Biol.* 1999;147(5):913-20.

Svensson, P. A., Jernås, M., Sjöholm, K., Hoffmann, J. M., Nilsson, B. E., Hansson, M., et al., Gene expression in human brown adipose tissue. *Int. J. Mol. Med.* 2011;27(2):227-32.

Takahashi, K., Okita, K., Nakagawa, M., Yamanaka, S. Induction of pluripotent stem cells from fibroblast cultures. *Nat. Protoc.* 2007;2(12): 3081-9.

Tamori, Y., Masugi, J., Nishino, N., Kasuga, M. Role of peroxisome proliferator-activated receptor-gamma in maintenance of the characteristics of mature 3T3-L1 adipocytes. *Diabetes.* 2002;51(7):2045-55.

Tchkonia, T., Giorgadze, N., Pirtskhalava, T., Tchoukalova, Y., Karagiannides, I., Forse, R. A., et al., Fat depot origin affects adipogenesis in primary cultured and cloned human preadipocytes. *Am. J. Physiol. Regul. Integr. Comp. Physiol.* 2002;282(5):R1286-96.

Thomas, K. R., Capecchi, M. R. Site-directed mutagenesis by gene targeting in mouse embryo-derived stem cells. *Cell.* 1987;51(3):503-12.

Toth, J. I., Yang, S. H., Qiao, X., Beigneux, A. P., Gelb, M. H., Moulson, C. L., et al., Blocking protein farnesyltransferase improves nuclear shape in fibroblasts from humans with progeroid syndromes. *Proc. Natl. Acad. Sci. US* 2005;102(36):12873-8.

Uldry, M., Yang, W., St-Pierre, J., Lin, J., Seale, P., Spiegelman, B. M. Complementary action of the PGC-1 coactivators in mitochondrial biogenesis and brown fat differentiation. *Cell Metab.* 2006;3(5):333-41.

Vantyghem, M. C., Vincent-Desplanques, D., Defrance-Faivre, F., Capeau, J., Fermon, C., Valat, A. S., et al., Fertility and obstetrical complications in women with LMNA-related familial partial lipodystrophy. *J. Clin. Endocrinol. Metab.* 2008;93(6):2223-9.

Varela, I., Pereira, S., Ugalde, A. P., Navarro, C. L., Suárez, M. F., Cau, P., et al., Combined treatment with statins and aminobisphosphonates extends longevity in a mouse model of human premature aging. *Nat. Med.* 2008; 14(7):767-72.

Varga, R., Eriksson, M., Erdos, M. R., Olive, M., Harten, I., Kolodgie, F., et al., Progressive vascular smooth muscle cell defects in a mouse model of Hutchinson-Gilford progeria syndrome. *Proc. Natl. Acad. Sci. US* 2006; 103(9):3250-5.

Vernochet, C., Azoulay, S., Duval, D., Guedj, R., Ailhaud, G., Dani, C. Differential effect of HIV protease inhibitors on adipogenesis: intracellular ritonavir is not sufficient to inhibit differentiation. *AIDS.* 2003;17(15):2177-80.

Verstraeten, V. L., Broers, J. L., Ramaekers, F. C., van Steensel, M. A. The nuclear envelope, a key structure in cellular integrity and gene expression. *Curr. Med. Chem.* 2007;14(11):1231-48.

Vigouroux, C., Caron-Debarle, M., Le Dour, C., Magré, J., Capeau, J. Molecular mechanisms of human lipodystrophies: from adipocyte lipid droplet to oxidative stress and lipotoxicity. *Int. J. Biochem. Cell Biol.* 2011;43(6):862-76.

Villarroya, F., Domingo, P., Giralt, M. Drug-induced lipotoxicity: lipodystrophy associated with HIV-1 infection and antiretroviral treatment. *Biochim. Biophys. Acta.* 2010;1801(3):392-9.

Wabitsch, M., Brüderlein, S., Melzner, I., Braun, M., Mechtersheimer, G., Möller, P. LiSa-2, a novel human liposarcoma cell line with a high capacity for terminal adipose differentiation. *Int. J. Cancer.* 2000;88(6): 889-94.

Waki, H., Tontonoz, P. Endocrine functions of adipose tissue. *Annu. Rev. Pathol.* 2007;2:31-56.

Weber, K., Plessmann, U., Traub, P. Maturation of nuclear lamin A involves a specific carboxy-terminal trimming, which removes the polyisoprenylation site from the precursor; implications for the structure of the nuclear lamina. *FEBS Lett.* 1989;257(2):411-4.

Wehnert, M., Muntoni, F. 60th ENMC International Workshop: non X-linked Emery-Dreifuss Muscular Dystrophy 5-7 June 1998, Naarden, The Netherlands. *Neuromuscul. Disord.* 1999;9(2):115-21.

Whyte, D. B., Kirschmeier, P., Hockenberry, T. N., Nunez-Oliva, I., James, L., Catino, J. J., et al., K- and N-Ras are geranylgeranylated in cells treated with farnesyl protein transferase inhibitors. *J. Biol. Chem.* 1997;272(22): 14459-64.

Wojtanik, K. M., Edgemon, K., Viswanadha, S., Lindsey, B., Haluzik, M., Chen, W., et al., The role of LMNA in adipose: a novel mouse model of lipodystrophy based on the Dunnigan-type familial partial lipodystrophy mutation. *J. Lipid Res.* 2009;50(6):1068-79.

Wu, J., Boström, P., Sparks, L. M., Ye, L., Choi, J. H., Giang, A. H., et al., Beige adipocytes are a distinct type of thermogenic fat cell in mouse and human. *Cell.* 2012;150(2):366-76.

Yang, S. H., Andres, D. A., Spielmann, H. P., Young, S. G., Fong, L. G. Progerin elicits disease phenotypes of progeria in mice whether or not it is farnesylated. *J. Clin. Invest.* 2008b;118(10):3291-300.

Yang, S. H., Bergo, M. O., Toth, J. I., Qiao, X., Hu, Y., Sandoval, S., et al., Blocking protein farnesyltransferase improves nuclear blebbing in mouse fibroblasts with a targeted Hutchinson-Gilford progeria syndrome mutation. *Proc. Natl. Acad. Sci. US* 2005;102(29):10291-6.

Yang, S. H., Chang, S. Y., Ren, S., Wang, Y., Andres, D. A., Spielmann, H. P., et al., Absence of progeria-like disease phenotypes in knock-in mice expressing a non-farnesylated version of progerin. *Hum. Mol. Genet.* 2011;20(3):436-44.

Yang, S. H., Meta, M., Qiao, X., Frost, D., Bauch, J., Coffinier, C., et al., A farnesyltransferase inhibitor improves disease phenotypes in mice with a Hutchinson-Gilford progeria syndrome mutation. *J. Clin. Invest.* 2006; 116(8):2115-21.

Yang, S. H., Qiao, X., Fong, L. G., Young, S. G. Treatment with a farnesyltransferase inhibitor improves survival in mice with a Hutchinson-Gilford progeria syndrome mutation. *Biochim. Biophys. Acta.* 2008a;1781 (1-2):36-9.

Yu, J., Vodyanik, M. A., Smuga-Otto, K., Antosiewicz-Bourget, J., Frane, J. L., Tian, S., Nie, J., Jonsdottir, G. A., Ruotti, V., Stewart, R., Slukvin, I. I., Thomson, J. A. Induced pluripotent stem cell lines *Science.* 2007 Dec. 21;318(5858):1917-20.

Zhang, F. L., Casey, P. J. Protein prenylation: molecular mechanisms and functional consequences. *Annu. Rev. Biochem.* 1996;65:241-69.

Zhang, J., Lian, Q., Zhu, G., Zhou, F., Sui, L., Tan, C., et al., A human iPSC model of Hutchinson Gilford Progeria reveals vascular smooth muscle and mesenchymal stem cell defects. *Cell Stem Cell.* 2011;8(1):31-45.

Zilberfarb, V., Piétri-Rouxel, F., Jockers, R., Krief, S., Delouis, C., Issad, T., et al., Human immortalized brown adipocytes express functional beta3-adrenoceptor coupled to lipolysis. *J. Cell Sci.* 1997;110 (Pt 7):801-7.

Zingaretti, M. C., Crosta, F., Vitali, A., Guerrieri, M., Frontini, A., Cannon, B., et al., The presence of UCP1 demonstrates that metabolically active adipose tissue in the neck of adult humans truly represents brown adipose tissue. *FASEB J.* 2009;23(9):3113-20.

Zuk, P. A., Zhu, M., Ashjian, P., De Ugarte, D. A., Huang, J. I., Mizuno, H., et al., Human adipose tissue is a source of multipotent stem cells. *Mol. Biol. Cell.* 2002;13(12):4279-95.

In: Adipogenesis
Editors: Y. Lin and X. Cai

ISBN: 978-1-62808-750-5

Chapter III

Adipogenesis and Osteoblastgenesis toward the Ageing Intervention in the 21st Century

Takeshi Imai*
Department Head of Aging Intervention
National Center for Geriatrics and Gerontology
Japan

Abstract

Population ageing is a major trend with global implications in the 21st century. Increasing Healthy Life Expectancy (HALE) is one of humanity's greatest achievements. Our recent study showed that a number of teeth of Japanese elderly people was associated with cognitive functions, and fat & bone parameters were decreasing in late elderly age. These data implicates the relation between adipogenesis and osteoblastgenesis to achieve increasing HALE. The adipocytes and osteoblasts are differentiated from the mesenchymal stem cells (MSC), suggesting that similar and different mechanisms are existed in MSC

* E-mail address: dai.ncgg@gmail.com.

differentiation into adipocytes and osteoblasts. Our final goal is how to expand HALE with regulating both adipogenesis and osteoblastgenesis. The major adipogenesis regulator is PPARγ, which is the member of nuclear receptor superfamily and ligand-dependent transcription factor. PPARγ-ablation in mature adipocyte leads to adipocyte death, lipodystrophy in mice. Runx2 is one of the osteoblastgenesis regulators. Runx2 mutation display cleidocranial dysplasia in human and mice, and Runx2 knock out mice showed embryonic lethality due to bone formation defects. The chemicals which regulate adipogenesis (PPARγ) and osteoblastgenesis (Runx2) will be discussed.

Abbreviations

15-deoxy-$\Delta^{12,14}$-PGJ_2; 15d-PGJ2
Blood sugar; BS
Hemoglobin; Hb
High density lipoprotein; HDL
Mesenchymal stem cell; MSC
Mini-Mental State Examination; MMSE
Peroxisome proliferator-activated receptor γ; PPARγ
Prostaglandin E_2; PGE2
Prostaglandin J_2; PGJ2
Prostaglandin J_2 Interacting Factor; PGJIF
Prostanoid; PG
Retinoid X receptors; RXRs
Total cholesterol; TC
Triglyceride; TG

1. Ageing in the World

Population ageing increases rapidly in the 21st century. Two persons celebrate their sixtieth birthday every second, and 58 million 60-years celebration in every year in the universe. So population ageing is progressing in all over the world, fastest in developing countries. Population ageing is the result of development of human society. Increasing Healthy Life Expectancy (HALE) is one of humanity's greatest achievements. People live longer because of improvement of nutrition, medical technique, health care, education and social & economic environment. Now the HALE of 33 counties is over 80 years. Japan is the only country whose population ageing over 60 years is more than 30%. My department of Ageing Intervention, of National Center for Geriatrics and Gerontology in Japan, therefore, was established to study evidence based increasing HALE. HALE - also called "disability-adjusted life expectancy" - represents the average number of years that a person can expect to live in "full health". This is calculated by taking into account years lived in less than full health due to disease and/or injury. The direct reasons of shortened HALE are diseases, such as obesity, metabolic syndrome, fracture, osteoporosis, cancer, dementia, *et ceteraet*c (UNFPA data).

2. Teeth and Cognitive Functions in Elderly People

A number of teeth of Japanese elderly people was associated with cognitive functions, lipid and sugar metabolism parameters. It has been reported that number of teeth is associated with cognitive function in elderly populations with dementia. However, little is known about this association in an ordinary elderly population. We evaluated this relationship in a Japanese population of elderly people aged. Dental examinations were performed all subjects with the Mini-Metal State Examination (MMSE) and Kohs task test for assessing cognitive function. Associations were not found between number of residual teeth and MMSE in total subjects or in males or females. However, associations were found between number of residual teeth and Kohs score in males. These results suggest that cognitive functions, especially, motor cognition, may be associated with number of teeth in ordinary elderly males (Imai et al., 2010a). Dementia is one of the most popular neurological disorders in the elderly. Some longitudinal studies have found significant associations between some life style-related diseases such as diabetes mellitus, hyperlipidemia, and dementia. Blood and dental examinations were performed for all subjects using the Mini-Mental State.

Examination (MMSE) for assessing cognitive function. Associations were not found between total cholesterol (TC), high density lipoprotein (HDL). triglyceride (TG), hemoglobin (Hb)-Alc, blood sugar (BS) in blood, and MMSE score in total subjects or in males or females. However, associations were found between the number of residual teeth and MMSE scores in both sexes. These results suggest that cognitive function may be associated with the number of teeth in ordinary elderly people. In addition, the number of residual teeth might be useful for predicting cognitive function in the elderly (Imai et al., 2010b).

3. Fat and Bone in Elderly People

To expand HALE, these physiological parameters above were re-calculated, surprisingly less number of parameters are deceased by age from 60 to 97 (data not shown). Among these Fat and bone values are similarly gradually declined by age (Figure 1, Imai 2006 and 2007, Arner et al., 2011, Spalding et al., 2008).

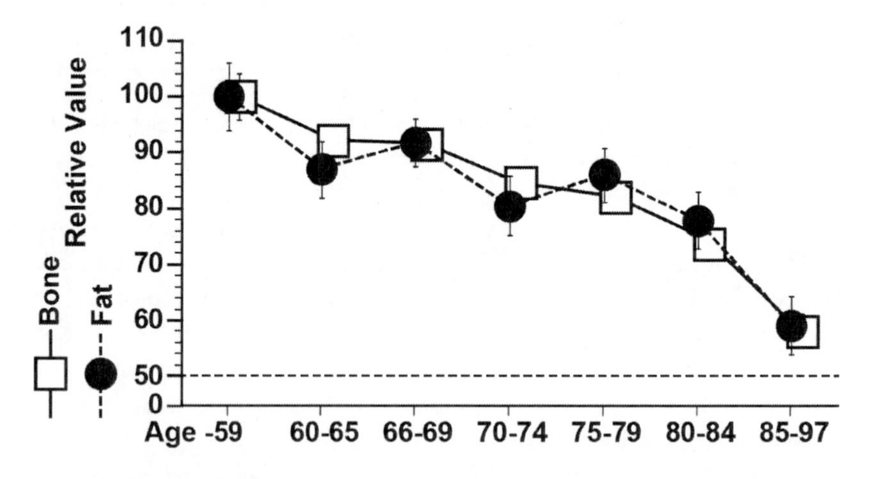

Figure 1. Bone and Fat in aged people.

Relative values of FAT and bone were represented in the graph. With comparison to younger age (below 60-year-old) control group, both values are ca 80% in 70-74 years old group, and ca 60 % in oldest group (85-97 year-old). Our data suggests adipocyte differentiation (adipogenesis) and osteoblast differentiation (osteoblastgenesis) are reduced with similar speed in elderly.

It suggested that similar mechanism existed in adipogenesis and osteoblastgenesis. Actually both cell types are differentiated from the mesenchymal stem cells. And some hormones, e.g. estrogen, are known to regulate both adipogenesis and osteoblastgenesis.

To increase the HALE, the common signaling pathways of regulating adipogenesis and osteoblastgenesis will be reviewed below.

4. Adipogenesis and Osteoblastgenesis from Mesenchymal Stem Cells

Mesenchymal stem cells (MSCs) are multipotent stromal cells that can differentiate into a variety of cell types, including: adipocyte, osteoblast, myocyte, chondrocyte and fibroblast. This phenomenon has been documented in specific cells and tissues in living animals and their counterparts growing in tissue culture (Figure 2 and Liu et al., 2012).

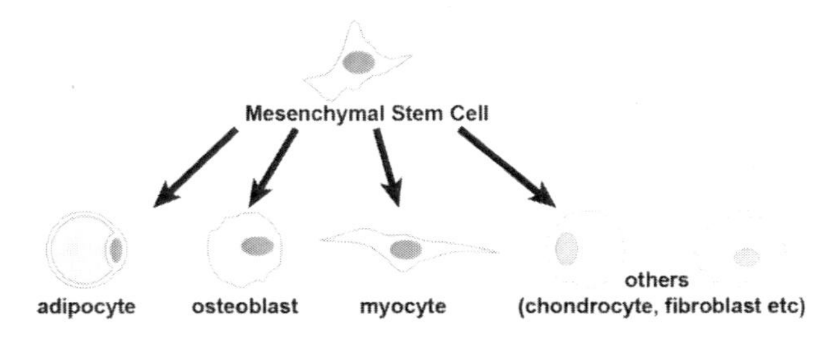

Figure 2. Mesenchymal Stem Cell (MSC).

5. Chemical-Regulated Adipogenesis and Osteoblastgenesis

It's indicating that the origin of adipocytes and osteoblasts were same cell type. In cell culture system, some cell lines have ability to differentiate into adipocyte and osteoblast after specific chemical treatment, suggesting that circumstances exposed with some hormones change their cell face from stem cells to adipocyte or osteoblast (Figure 3).

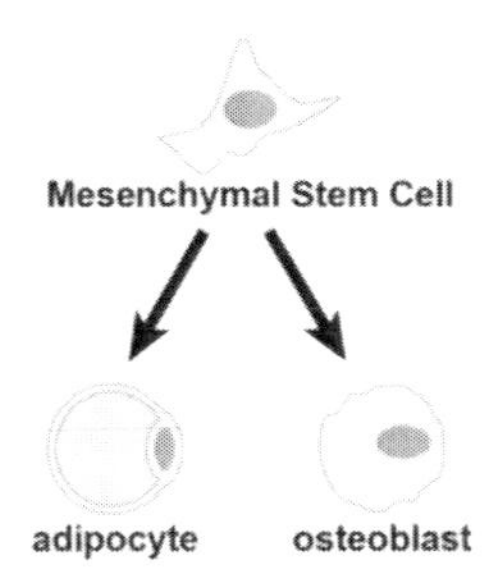

Figure 3. Adipogenesis and osteoblastgenesis from mesenchymal stem cell.

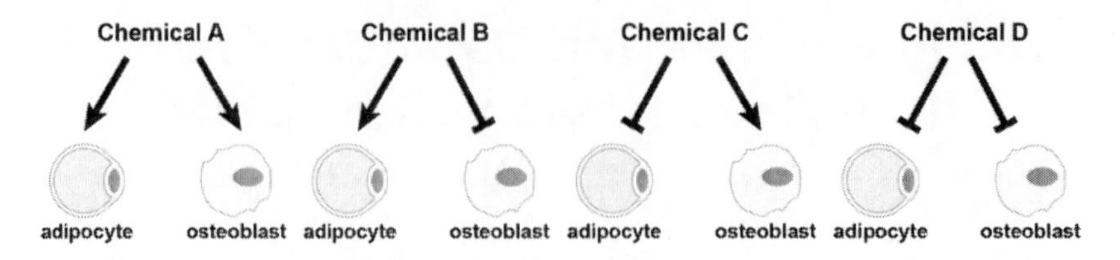

Figure 4. Chemical regulated adipogenesis and osteoblastgenesis.

Actually some chemicals regulate adipogenesis and osteoblastgenesis. The chemicals were classified into 4 groups (Figure 4).

In elderly, adipogenesis and osteoblastgenesis behave similarly reduced (Figure 1), indicating that chemical A or chemical D in figure 4 are our interests to expand the HALE.

6. PPARγ in Adipocyte and Runx2 in Osteoblast

How chemical A and D regulate both adipogenesis and osteoblastgenesis in same direction? Most probably the chemical target proteins (so called receptors) are peroxisome proliferator-activated receptor γ (PPARγ) in adipocytes and Runx2 in osteoblasts.

The nuclear receptor PPARγ is a ligand-dependent transcriptional factor that heterodimerizes with retinoid X receptors (RXRs) and is activated by natural ligands, such as arachidonic acid metabolites, prostanoids (PGs) and fatty acid-derived components, and the insulin-sensitizing drugs thiazolidinediones (TZD). Activation of PPARγ by TZD in white and brown preadipocyte cell lines results in robust differentiation into adipocytes, and TZD administration to rodents increases accumulation of white (WAT) and brown (BAT) adipose tissue deposits. Moreover, overexpression of PPARγ in fibroblasts induces adipogenesis, and PPARγ-null embryonic stem cells and fibroblastic cells from PPARγ-deficient embryos cannot differentiate into adipocytes in vitro. PPARγ is also known to be indispensable for adipose tissue formation in vivo, because mice chimeric for WT and PPARγ-null cells show little or no contribution of null cells to adipose tissue, and PPARγ-deficient pups, derived by tetraploid rescue that bypasses placental defects, lack BAT and WAT (Imai et al., 2004). To study PPARγ functions in mature adipocytes, we ablated PPARγ in adipocytes of adult mice through conditional somatic mutated systems (Cre-ER^{T2}) selectively in brown and white adipocytes. PPARγ-deficient adipocytes die within a few days, which triggers an inflammatory reaction in BAT and WAT, and are replaced by newly differentiated PPARγ-expressing adipocytes, which most probably derive from fibroblast-like preadipocyte cells (Imai 2003, and Imai et al., 2001 and 2004). TZD induced adipogenesis and obesity through binding to PPARγ (Metzger et al., 2005).

Osteoblast transcription factor Runx2 was identified as transcription factor for controlling osteocalcin transcription (Ducy et al., 1997).

Runx2 genetic ablated mice displayed bone abnormality, embryonic lethality (Komori et al., 1997; Mundlos et al., 1997; Otto et al., 1997; Rodan and Harada 1997). And heterozygous mice showed similar phenotype to Cleido-cranial dysostosis (CCD). A human point genetic mutation in Runx2 locus leads to CCD (Mundlos et al., 1997; Otto et al., 1997; Rodan and Harada 1997). Taken together Runx2 behave osteogenic transcription factor.

The chemicals which regulate PPARγ and Runx2 activities are discussed below.

7. Drug (TZD) Regulated Adipogenesis and Osteoblastgenesis

TZD is drug for insulin sensitizer. Molecular action of TZD is supposed to bind to PPARγ with very high affinity described above. Moreover, one of the side effects of TZD is decreasing bone marrow and BMI (bone mineral density). Molecular mechanism is not improved, but the possible reason is as follows;

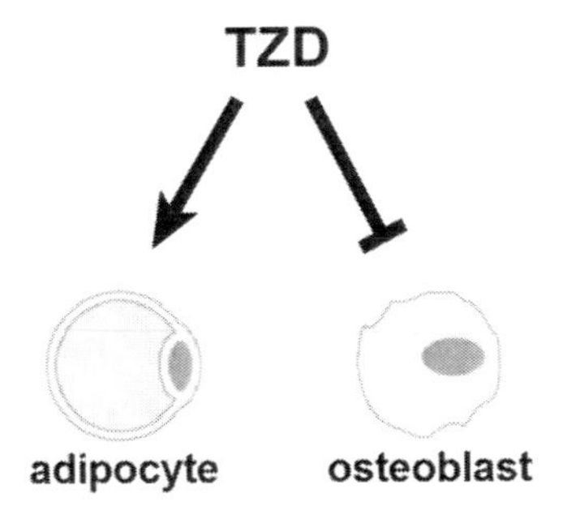

Figure 5. TZD regulated adipogenesis and osteoblastgenesis.

Both adipocytes and osteoblasts are differentiated from same precursor, mesenchymal stem cells. TZD bind and activates PPARγ and induced strongly to adipocyte differentiation from mesenchymal stem cells, suggesting that the most of mesenchymal stem cells differentiated into adipocytes and less number of mesenchymal stem cells are exists (left) in the bone marrows. Using stem cell line experiment it was proved *in vitro* (data not shown). Taken together TZD bind to PPARγ and induced adipogenesis directly, but reduced osteoblastgenesis indirectly with unknown mechanism, suggesting that TZD is the chemical type B, not A or D.

8. Estrogen Effect on Both Adipogenesis and Osteoblastgenesis

Estrogens are a group of compounds named for their importance in both menstrual and estrous reproductive cycles. Most of the osteoporosis in aged people is caused by menopause, lacking estrogen induces high turnover type of osteoporosis, indicated that estrogen repressed osteoblastgenesis.

As we know, fat has gender differences, and previous mice genetics study showed that estrogen regulated adipogenesis similar to PPARγ (Figure 4).

Estrogen is also chemical B, which is different for elderly people.

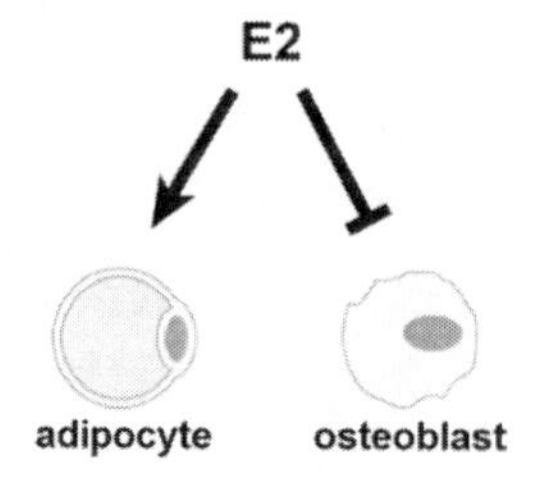

Figure 6. E2 effect on the adipogenesis and osteoblastgenesis.

9. Prostaglandin J2 (PGJ2) Regulate Both Adipogenesis and Osteoblastgenesis

Other hormones, prostanoids (PGs) regulated adipogenesis described previously. PGs bind and activate PPARγ. Among PG family, PGJ2 derivatives bind to PPARγ with high affinity (Imai et al., 2004), especially 15-deoxy-$\Delta^{12,14}$-PGJ_2 (15d-PGJ2).

PGJ2s also regulated osteoblastgenesis on transcriptional levels (manuscript in preparation). In osteoblast cell culture system PGJ2 reduced Rhodamine 123 incorporation in mitochondria (manuscript in preparation). PGJ2 is also chemical B, which is different for elderly people.

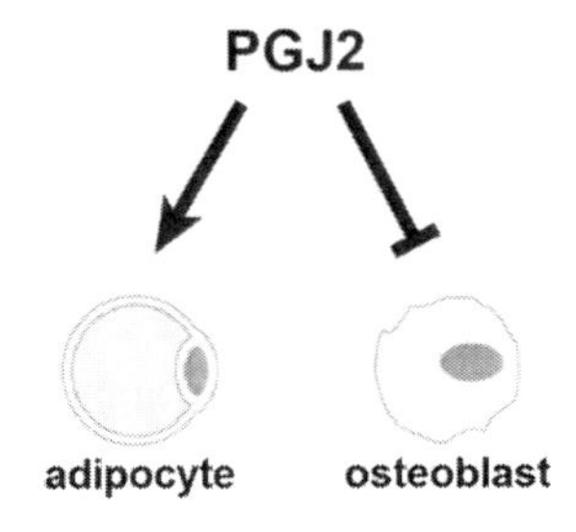

Figure 7. PGJ2 effect on the adipogenesis and osteoblastgenesis.

10. Prostaglandin X2 (PGX2) Regulate Both Adipogenesis and Osteoblastgenesis

Using nanobeads technology, PGJ2 target protein complex was purified and the components were identified (manuscript in preparation). This PGJ2 target protein bound to PGJ2 directly, and bound to other prostaglandins, PGX2 (manuscript in preparation). Surprisingly PGX2 stimulated osteoblastgenesis as well as adipogenesis, suggesting that PGX2 is the chemical which activated both adipogenesis and osteoblastgenesis. Most probably PGX2 concentration is decreasing by age, and achieved less amount of fat and bone in elderly people.

11. Nanotechnology for Identification of Chemical Target Protein

11.1. Arginine

Arginine exhibits a wide range of biological activities through a complex and highly regulated set of pathways that remain incompletely understood at both the whole-body and the cellular levels. Therefore effective purification system for arginine interacting factors (AIFs) was recently developed with novel magnetic nanobeads (FG beads) composed of magnetite particles/glycidyl methacrylate (GMA)–styrene copolymer/covered GMA. These nanobeads have shown higher performance compared with commercially available magnetic beads in terms of purification efficiency. We have newly developed L-arginine methyl ester (L-AME)-immobilized beads by conjugating L-AME to the surface of these nanobeads. At first inducible nitric oxide synthase, which binds and uses L-arginine as a substrate, specifically bound to L-AME-immobilized beads. Secondly, we newly identified phosphofructokinase, RuvB-like 1 and RuvB-like 2 as AIFs from crude extracts of HeLa cells using nanobeads. The data presented here demonstrate that L-AME-immobilized beads are effective tool for purification of AIFs directly from crude cell extracts. We expect that the present method can be used to purify AIFs from various types of cells (Hiramoto et al., 2010, Hotta et al., 2013).

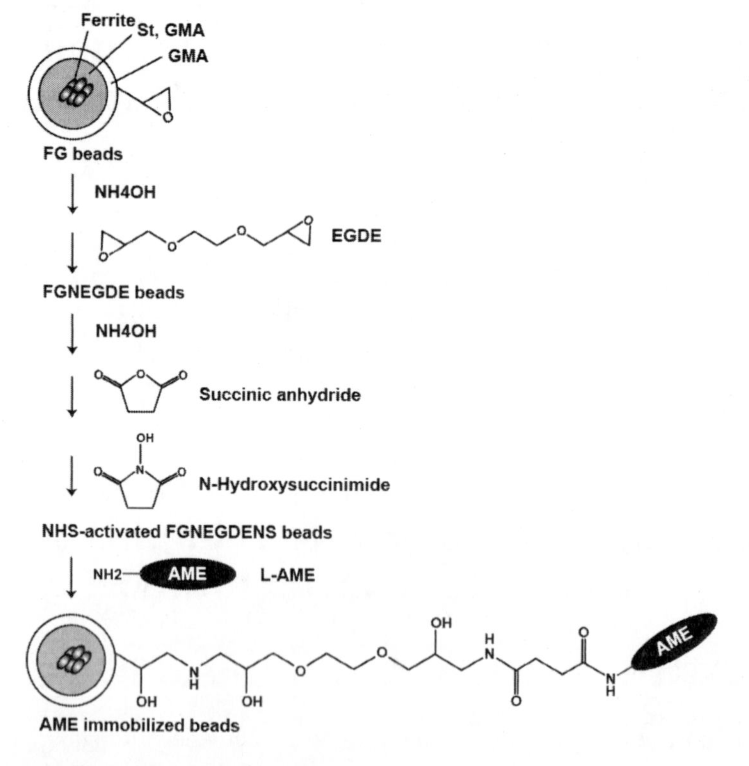

Figure 8. AME-immobilized nanobeads.

11.2. Prostaglandin J_2 (PGJ2)

Prostaglandin J_2 (PGJ2) family have been reported to show various kinds of biological activities. Considerable progress has been made toward understanding the mechanism of adipogenesis, however, the mechanisms of other actions of PGJ2 family remain controversial. The 15d-PGJ2 is one of the members of PGJ2 family, and is known as a ligand for PPARγ, which promotes the expression of the crucial genes for adipogenesis. Maekawa et al., 2011 found that 15d-PGJ2 did not stimulate PPARγ-mediated gene expression in HEK293 cells whereas 15d-PGJ2 transactivated PPARγ-dependent transcription in other cell lines. Moreover, we confirmed that 15d-PGJ2 suppressed the growth of HEK293 cells. These observations suggest that 15d-PGJ2 shows another biological activity e.g. growth inhibition in HEK293 cells via unknown receptor for 15d-PGJ2.

The aim of this study is to develop and validate effective purification system for PGJ2 interacting factors (PGJIFs). We have recently developed high performance magnetic nanobeads. In this study, we have newly developed 15d-PGJ2-immobilized beads by conjugating 15d-PGJ2 to the surface of these nanobeads.

Firstly, we showed that PPARγ specifically bound to 15d-PGJ2-immobilized beads. Secondly, we newly purified and identified new PGJIF from crude extracts of HEK293 cells using this affinity purification system. These data presented here demonstrate that 15d-PGJ2-immobilized beads are effective tool for purification of PGJIFs directly from crude cell extracts (Maekawa et al., 2011 and Karasawa et al., 2013).

Conclusion

To expand HALE, chemical (hormone), which regulate both PPARγ in adipocytes and Runx2 in osteoblasts, was possibly identified, prostanoids (PGX2). The PGX2 binds to PPARγ in adipocytes, and activated adipogenesis. In osteoblast, PGX2 binds to target protein and activated Runx2 activity in the nuclear.

References

Arner P, Bernard S, Salehpour M, Possnert G, Liebl J, Steiner P, Buchhotz BA, Eriksson M, Arner E, Hauner H, Skurk T, Ryden M. Frayn KN & Spaiding KL. Dynamics of human adipose lipid turnover in health and metabolic disease. *Nature* 478:110-113, 2011.

Ducy P, Zhang R, Geoffroy V, Ridall AL, Karsenty G. Osf2/Cbfa1: A Transcriptional Activator of Osteoblast Differentiation. *Cell* 89:747-754, 1997.

Hiramoto M, Maekawa N, Masaike Y, Kuge T, Ayabe F, Watanabe A, Masaike Y, Hatakeyama M, Handa H & Imai T*. High-performance affinity chromatography method for identification of L-arginine interacting factors using magnetic nanobeads. *Biomed Chromatogr* 24:606-612, 2010.

Hotta K, Nashimoto A, Yasumura E, Suzuki M, Azuma M, Shima D, Nabeshima R, Hiramoto M, Okada A, Sakata-Sogawa K, Tokunaga M, Ito T, Sakamoto S, Kabe Y, Aizawa S, Imai T, Yamaguchi Y, Watanabe H & Handa H. Vesnarinone suppresses TNF□ mRNA expression by inhibiting valosin-containing proteins. *Mol Pharm* 83:930-938, 2013.

Imai T, Jiang M, Chambon P & Metzger D. Impaired adipogenesis and lipolysis in the mouse upon Cre-ERT2-mediated selective ablation of RXR alpha in adipocytes. *Proc Natl Acad Sci USA* 98:224-228, 2001.

Imai T, Nishinaga M & Matsushita K. Association with the cognitive functions, the number of teeth and cognitive function in elderly population. *Kagoshima* 61:47-51, 2010. In Japanese.

Imai T, Nishinaga M, Nakamura T, Okumiya K, Matsubayashi K, Doi Y & Matsushita K. Association with the lipid and sugar metabolism parameters, the number of teeth and cognitive function in elderly population. *Aichigakuin* 48:59-66, 2010. In Japanese.

Imai T, Takakuwa R, Marchand S, Dentz E, Bornert JM, Messaddeq N, Wendling O, Mark M, Desvergne B, Wahli W, Chambon P & Metzger D. PPARγ is required in mature white and brown adipocytes for their survival in the mouse. *Proc Natl Acad Sci USA* 101:4543-4547, 2004.

Imai T. *Aging Intervention*. Iryo 60:29-32, 2006. In Japanese.

Imai T. *Evidence based anti*-aging. Keio Igaku 84:149-155, 2007. In Japanese.

Imai T. Functional genetic dissection of nuclear receptor signaling in obesity, diabetes and liver regeneration using spatio-temporally controlled somatic mutagenesis in the mouse. *Keio Journal of Me*dicine 52: 198-203, 2003.

Karasawa S, Azuma M, Kasama T, Sakamoto S, Kabe Y, Imai T, Yamaguchi Y, Miyazawa K & Handa H. Vitamin K2 covalently binds to Bak and induces Bak-mediated apoptosis. *Mol Pharm* 83:613-620, 2013.

Komori T, Yagi H, Nomura S, Yamaguchi A, Sasaki K, Deguchi K, Shimizu Y, Bronson RT, Gao YH, Inada M, Sato M, Okamoto R, Kitamura Y, Yoshiki S, Kishimoto T. Targeted Disruption of Cbfa1 Results in a Complete Lack of Bone Formation owing to Maturational Arrest of Osteoblasts. *Cell* 89:755-764, 1997.

Liu Y, Berendsen AD, Jia S, Lotinun S, Baron R, Ferrara N & Olsen B. Intracellular VEGF regulates the balance between osteoblast and adipocyte differentiation. *J Clin Invest* 122:3101-3111, 2012.

Maekawa N, Hiramoto M, Sakamoto S, Ikeda M, Naitou M, Acharya HP, Kobayashi Y, Suematsu M, Handa H & Imai T*. High performance affinity chromatography method for identification of 15-deoxy-Δ12,14-prostaglandin J2 interacting factors using magnetic nanobeads. *Biomed Chromatogr* 25:466-471, 2011.

Metzger D, Imai T, Jiang M, Takakuwa R, Desvergne B, Wahli W & Chambon P. Functional role of RXRs and PPAR□ in mature adipocytes. Prostaglandins, *Leukot Essent Fatty Acids* 73:51-58, 2005.

Mundlos S, Otto F, Mundlos C, Mulliken JB, Aysworth AS, Albright S, Lindhout D, Cole WG, Henn W, Knoll JHM, Owen MJ, Mertelsmann R, Zabel BU, Olsen BR. Mutations Involving the Transcription Factor CBFA1 Cause Cleidocranial Dysplasia. *Cell* 89:773-779, 1997.

Otto F, Thornell AP, Crompton T, Denzel A, Gilmour KC, Rosewell IR, Stamp GWH, Beddington RSP, Mundlos S, Olsen BR, Selby PB, Owen MJ. Cbfa1, a Candidate Gene for Cleidocranial Dysplasia Syndrome, Is Essential for Osteoblast Differentiation and Bone Development. *Cell* 89:765-771, 1997.

Rodan GA and Harada S. The missing bone. *Cell* 89:677-680, 1997.

Spalding KL, Arner E, Westermark, Bernard S, Buchholz BA, Bergmann O, Blomqvist L, Hoffstedt J, Naslund E, Britton T, Concha H, Hassan M, Ryden M, Frisen J & Arner P. Dynamics of fat cell turnover in humans. *Nature* 453:783-785, 2008.

UNFPA (United Nations Population Fund) data. Ageing in the twenty-first century: a celebration and a challenge. Executive Summary. Ageing in the Twenty-First Century: A Celebration and A Challenge. Published by the United Nations Population Fund. (UNFPA), New York, and HelpAge International, London. Copyright © United Nations Population Fund (UNFPA) and HelpAge International, 2012. All rights reserved.

In: Adipogenesis
Editors: Y. Lin and X. Cai

ISBN: 978-1-62808-750-5

Chapter IV

Role of Wnt/β-Catenin Signaling in Bone Marrow adiposity Following Cancer Chemotherapy

Kristen R. Georgiou and Cory J. Xian [*]
Sansom Institute for Health Research,
School of Pharmacy and Medical Sciences, University of South Australia,
Adelaide, SA, Australia

Abstract

The bone marrow microenvironment is home to haematopoietic and mesenchymal cell populations that regulate bone turnover through complex interactions. The high proliferative capacity of these cell populations makes them susceptible to damage and injury, which alters the steady-state function of the bone marrow environment. Following cancer chemotherapy, irradiation and long-term glucocorticoid use, a fatty marrow cavity is typically observed, whereby reduced bone and increased fat formation of marrow stromal progenitor cells often results in increased marrow fat, reduced bone mass and increased fracture risk. Although the underlying mechanisms remain to be clearly elucidated, recent investigations have suggested a switch in lineage commitment of

[*] Correspondence to CJ Xian: cory.xian@unisa.edu.au.

bone marrow mesenchymal stem cells down the adipogenic lineage at the expense of osteogenic differentiation, following damage caused by treatment regimens. As the Wnt/β-catenin signaling pathway has been recognized as the key mechanism regulating stromal commitment, its involvement in the osteogenic and adipogenic lineage commitment switch under damaging conditions has been of great interest. This chapter will review the effects of chemotherapy treatment regimens on commitment to the adipogenic and osteogenic lineages of bone marrow stromal progenitor cells. It will also summarize the Wnt/β-catenin signaling pathway and its role in stromal cell lineage commitment and recovery after damage, as well as its potential use as a therapeutic target.

I. Introduction

The bone marrow is home to mesenchymal and haematopoietic cell lineages, whereby a regulated balance of proliferation and differentiation of these cell types is required for a steady-state functioning marrow. This includes bone formation from mesenchymal stem cells (MSCs, also referred to as bone marrow stromal stem cells or BMSCs) and haematopoiesis from haematopoietic stem cells (HSCs). Damaging conditions, such as those observed following cancer treatments, disrupt the homeostatic balance of cell populations within the bone marrow. Intensive, long-term cancer chemotherapy or radiotherapy treatments are associated with depletion of the haematopoietic fraction of the bone marrow, as well as altered differentiation of stromal progenitor cells. Damage to the bone marrow microenvironment observed following short-term treatments or with low doses of chemotherapy or radiation, is transient and re-establishment of bone marrow cell populations is observed. However, despite this potential for recovery, following repeated insults as in treatment regimens, stem cell populations that enable re-establishment of the marrow are induced to differentiate and over time diminish, thus recovery becomes less efficient. Such changes to the marrow are typically associated with bone loss and an adipocyte-rich phenotype, which increases fracture risk.

In order to gain a better understanding of the underlying pathobiology and to explore potential means of enhancing recovery, it is of great interest to unravel the mechanisms that are associated with a switch in bone and fat lineage commitment and differentiation. The Wnt/β-catenin signaling pathway is one such regulatory mechanism of interest. The Wnt/β-catenin signaling pathway is a known regulator of mesenchymal lineage commitment, thus its involvement in the potential commitment switch between osteogenic and adipogenic lineages is of great interest and one that requires further attention. This chapter aims to illustrate the altered cellular composition of the bone marrow cavity following damaging cancer chemotherapy and glucocorticoid therapy, focusing on the role of the Wnt/β-catenin signaling pathway in the regulation of stromal cell lineage commitment and differentiation.

II. Osteogenesis, Adipogenesis, and Regulation by Wnt/β-Catenin Signaling Pathway

BMSCs have the capacity to differentiate into multiple cell lineages including osteoblasts, adipocytes, chondrocytes and myocytes (Peled, Petit et al., 1999). The commitment of MSCs to a particular lineage is largely dependent on transcription factors and complex signaling mechanisms within the marrow microenvironment. Enabling this are close interactions between cells of the mesenchymal and haematopoietic lineages, which support appropriate stromal commitment and haematopoiesis.

Osteogenesis and Osteogenic Transcription Factors

Along the osteogenic cell lineage, BMSCs are firstly committed to become highly proliferative osteoprogenitor cells which then develop into pre-osteoblasts and further differentiate into mature osteoblasts (Long 2001). Osteogenesis is a complex process, initiated and regulated by transcription factors runt-related transcription factor 2 (Runx2) and osterix (Osx), whose expression is regulated by β-catenin and homeobox protein MSX2 and repressed by adipogenic transcriptional factor peroxisome proliferator activated receptor-gamma (PPAR-γ) (Kitazawa, Mori et al., 2008; Scheideler, Elabd et al., 2008). In the early stages of differentiation, osteoblasts synthesise an extracellular matrix that consists primarily of type I collagen and go on to express alkaline phosphatase (ALP), indicative of the osteoblast phenotype. The complex of transcription factors go on to activate target genes responsible for osteoblast maturation, such as bone sialoprotein and osteocalcin, resulting in mineralisation of bone (Kitazawa, Mori et al., 2008). At the end stage, osteoblasts actively synthesise and mineralise the bone matrix (Chaudhary, A.

et al., 2004) and eventually become bone lining cells or osteocytes embedded within bone matrix which regulate remodelling and maintenance of the trabecular bone structure and strength (Bonewald and Johnson 2008). There are a number of contributing factors associated with the commitment and differentiation of BMSCs, one of particular note is the Wnt/β-catenin signaling pathway, which will be discussed in more details below.

Adipogenesis and Adipogenic Transcription Factors and Regulators

Aging, menopause and medical treatments including glucocorticoid therapy, irradiation or chemotherapy, have all been associated with bone loss and marrow adiposity (Moerman, Teng et al., 2004; Li, Luo et al., 2008; Muruganandan, Roman et al., 2009; Georgiou, Hui et al., 2012; Georgiou, Scherer et al., 2012). As adipocytes and osteoblasts share the MSC as a precursor, it has been proposed that differentiation down the adipogenic program is preferential to osteogenesis under such conditions. The tightly regulated process of adipogenesis is enabled by a cascade of transcription factors, resulting in the formation of mature adipocytes (Scheideler, Elabd et al., 2008). Transcription factors PPAR-γ and CCAAT-enhancer binding protein-alpha (C/EBPα) form a complex that goes on to regulate the expression of adipogenic genes, which enable adipocyte commitment and terminal differentiation. In this instance, glycerol-3 phosphate dehydrogenase, hormone-sensitive lipase, fatty acid synthase, fatty acid binding proteins (FABPs) and perilipin are induced, as well as the secretion of adipokines such as leptin, adiponectin, adipsin, tumor necrosis factor (TNF-α) and retinol binding protein 4 (Scheideler, Elabd et al., 2008).

Wnt Signaling Pathway in Osteogenic and Adipogenic Commitment and Differentiation

The transcriptional regulation of the balance between osteoblast and adipocyte differentiation is a complex one and a clearer understanding is required. The canonical Wnt signaling pathway or the Wnt/β-catenin signaling pathway is well established as an essential regulator of stromal commitment, maintaining osteogenic differentiation, whilst suppressing excess adipogenesis. As a brief overview, the Wnt/β-catenin signaling pathway is activated upon a Wnt ligand binding to the frizzled (Fzd) and lipoprotein-related protein 5/6 (LRP5/6) co-receptor complex, which recruits an intracellular molecule disheveled (Dsh), inhibiting the action of a complex consisting of intracellular molecules that act to phosphorylate and tag β-catenin for ubiquitin-mediated degradation. When this is inhibited, β-catenin accumulates in the cytoplasm and translocates to the nucleus to elicit target gene transcription (Macsai, Foster et al., 2008).

Canonical Wnt ligands include Wnt1, Wnt3a, Wnt7b, and Wnt10b, which when bound to the LRP5/6 co-receptor complex activate the Wnt/β-catenin signaling pathway. Wnt10b, expressed by pre-adipocytes and stromal vascular cells, has been illustrated in a number of studies to be associated with regulating the balance between osteogenesis and adipogenesis, stimulating osteogenic differentiation and inhibiting adipogenic differentiation by maintaining pre-adipocytes in an undifferentiated state (Bennett, Longo et al., 2005; Macsai, Foster et al., 2008). It does so by suppressing PPAR-γ and C/EBPα expression, causing inhibition of terminal adipogenic differentiation and inducing Runx2, Dlx5 and Osx to promote osteogenesis (Ross, Hemati et al., 2000; Bennett, Longo et al., 2005; Macsai, Foster et al., 2008). The role and importance of Wnt10b and the Wnt/β-catenin signaling pathway in regulating adipogenic commitment and differentiation has been further characterized in recent studies (Zhou, Mak et al., 2008; Cawthorn, Bree et al., 2012; Chung, Lee et al., 2012). Interestingly, a link between Wnt signaling and adipogenic commitment and differentiation was demonstrated, where C/EBPβ binds directly to the Wnt10b promoter, inhibiting Wnt10b expression over the course of adipogenic differentiation. Wnt10b was also found to enhance osteogenic differentiation of stromal cells and dampen pre-adipocyte differentiation in a β-catenin-dependent manner as the Wnt10b anti-adipogenic effect was found suppressed in a β-catenin knockdown model (Cawthorn, Bree et al., 2012; Chung, Lee et al., 2012).

There are a number of secreted Wnt antagonists that bind to Wnt ligands or co-receptors, preventing Wnt/β-catenin signaling activation. In this instance, β-catenin is degraded in the cytoplasm, preventing or dampening target gene transcription. Notable antagonists include secreted frizzled-related protein-1 (sFRP-1), sclerostin, dickkopf-1 (Dkk-1) and Wnt inhibitor factor-1 (Wif-1) (Macsai, Foster et al., 2008; Naito, Omoteyama et al., 2012). The potential role of deregulation of the Wnt/β-catenin signaling pathway by the above antagonists has been investigated in relation to the increased marrow adiposity and bone loss observed in osteoporosis. Interestingly, osteoporotic women have higher serum Dkk-1 than controls, thought to be influencing

Wnt/β-catenin signaling and thus the regulation of bone formation (Anastasilakis, Polyzos et al., 2010).

The Wnt antagonist sFRP-1 has also been suggested as a player in deregulated osteoblast and adipocyte commitment and differentiation. It has been shown to be secreted by pre-adipocytes and negatively regulates osteogenic differentiation potential *in vitro* (Taipaleenmaki, Abdallah et al., 2011; Abdallah and Kassem 2012). Pre-adipocytes, osteoblasts and pre-osteocytes contribute to sFRP-1 production and thus regulation of Wnt/β-catenin signaling in this manner (Bodine, Billiard et al., 2005). Over expression of sFRP-1 in osteoblast and pre-osteocyte cell lines increased apoptosis and antagonism of Wnt/β-catenin signaling (Bodine, Billiard et al., 2005). Transcription of the Wnt inhibitor sFRP-1 is regulated by Gli1 and Gli2 as a consequence of increased Hedgehog (Hh) signaling (He, Sheng et al., 2006), suggesting existence of a molecular link between Hh and Wnt signaling pathways and demonstrating an extensive signaling network that regulates bone and bone marrow maintenance throughout life.

III. Potential Roles of Wnt/β-Catenin Signaling Pathway in Cancer Chemotherapy- and Glucocorticoid Therapy-Induced Bone Loss and Marrow Adiposity

Cancer Chemotherapy-Induced Bone Loss and Marrow Adiposity

Cancer chemotherapy treatment is associated with an altered balance of cell populations within the bone marrow cavity, including haematopoietic stem cells, stromal stem cells and their respective progeny. Myelosuppression, or marrow haematopoietic ablation is a typical consequence of intensive cancer therapy in patients as well as in animal studies (Stevens, Moore et al., 1990; Fogelman, Blake et al., 2003; Michal, Adelstein et al., 2012; Poncin, Beaulieu et al., 2012). Such defects result in altered cellular composition of the bone marrow such as an increased fatty marrow cavity, which has effects on steady-state function. Adipocytes have been illustrated to have a negative effect on haematopoiesis, whereby an adipocyte-rich marrow reduces not only haematopoietic progenitor cell number and differentiation, but also affects HSC cell cycling (Belaid-Choucair, Lepelletier et al., 2008; Naveiras, Nardi et al., 2009). Thus interactions between each cell type are vital to an adequately functioning marrow environment and therefore any disturbance to either cell type results in marrow dysfunction (Bianco 2011; Despars and St-Pierre 2011).

Long-term cancer chemotherapy disrupts the complex network of signaling pathways that regulate lineage commitment, differentiation, and balance within the marrow cavity (Georgiou, Foster et al., 2010; Fan, Georgiou et al., 2011). In the clinic, patients receiving chemotherapy and irradiation treatment for Hodgkins disease, seminoma, prostate and breast cancer have an observed increase in central marrow fat (Stevens, Moore et al., 1990; Fogelman, Blake et al., 2003), however mechanisms for these are largely unclear. In order to develop specific and targeted therapies to combat cancer treatment-induced bone and bone marrow defects, the cellular and molecular mechanisms that are responsible for the changes in lineage determination must be elucidated.

Direct exposure of MSCs to chemotherapeutic agents doxorubicin and etoposide *in vitro* resulted in a preferential differentiation capacity towards the adipogenic lineage at the expense of the osteogenic lineage (Buttiglieri, Ruella et al., 2011). In rat models of methotrexate and 5-fluorouricil (5-FU) chemotherapy, a reduction in trabecular bone volume and increased marrow adiposity are apparent over the treatment time-course (Xian, Cool et al., 2006; Xian, Cool et al., 2008; Fan, Cool et al., 2009; Georgiou, King et al., 2012; Georgiou, Scherer et al., 2012; King, Georgiou et al., 2012; Raghu Nadhanan, Abimosleh et al., 2012). Furthermore, an increased adipogenic differentiation potential at the expense of osteogenic differentiation, is observed in *ex vivo* cultured BMSCs of methotrexate or 5-FU-treated rats (Georgiou, Scherer et al., 2012; Raghu Nadhanan, Abimosleh et al., 2012). Recently, this switch in lineage commitment was at least partly attributed to deregulation of the Wnt/β-catenin signaling pathway, whereby the methotrexate chemotherapy-induced increase in marrow adiposity and bone loss were alleviated by treatment with an agonist of the Wnt/β-catenin pathway (Georgiou, King et al., 2012). Interestingly, expression of both Dkk-1 and sFRP-1 mRNA was found up-regulated one day after completion of 5-days of methotrexate treatment (day 6) and sFRP-1 mRNA expression was found to be elevated 4 days after completion of treatment (day 9), illustrating the involvement of Wnt antagonists in deregulation of Wnt signaling in this model (Georgiou, King et al., 2012).

Glucocorticoid Therapy-Induced Bone Loss and Marrow Adiposity

Long-term therapeutic glucocorticoid use has been established as a risk factor for osteoporosis and increased fracture risk, particularly associated with its effects on bone formation (Saag, Shane et al., 2007). In both human and rat cell culture studies, dexamethasone dose-dependently increases adipogenic differentiation of BMSCs and adipogenic-associated gene expression, with a corresponding reduction in the expression of osteogenic transcription factors Runx2 and Osx (Kitajima, Shigematsu et al., 2007; Hung, Yeh et al., 2008; Lin, Dai et al., 2010). The Wnt/β-catenin signaling pathway has been established as a regulator, since glucocorticoid-induced up-regulation of Wnt signaling antagonists Dkk-1, sclerostin and Wif-1 in animal, cell culture and clinical studies has been observed (Yao, Cheng et al., 2008; Naito, Omoteyama et al., 2012). As such, serum from patients undergoing glucocorticoid therapy containing high levels of Dkk-1 directly reduced osteoblast differentiation in *in vitro* culture and exposure of primary osteoblasts to dexamethasone caused a reduction in ALP activity, as well as reducing cytoplasmic and nuclear β-catenin (Butler, Queally et al., 2010; Brunetti, Faienza et al., 2013). Silencing Dkk-1 in these osteoblasts alleviated the reduction in ALP activity and in a corresponding study, Dkk-1 knockdown resulted in abrogation of the observed glucocorticoid-mediated increase in adipogenic differentiation potential (Wang, Ko et al., 2008; Butler, Queally et al., 2010). Furthermore, in *in vitro* cell culture, dexamethasone-induced adipogenesis was found to be correlated with dampened Wnt/β-catenin signaling, increased levels of antagonists Dkk-1 and Wif-1, increased protein expression of GSK-3β and a corresponding reduction in β-catenin protein. In addition, β-catenin knockdown alone was found to enhance the dexamethasone-induced adipogenic differentiation and dampen Wnt/β-catenin signaling (Naito, Omoteyama et al., 2012). Similarly, it has been shown that antagonist sFRP-1 contributes to glucocorticoid-induced disruption of Wnt/β-catenin signaling *in vitro*. Supplementation of dexamethasone in MC3T3 osteoblast culture caused a significant reduction in MC3T3 mineralization potential, associated with increased sFRP-1 mRNA expression, and a reduction in β-catenin (Hayashi et al. 2008). These studies illustrate that glucocorticoid-induced effects on osteogenic differentiation are at least in part a result of attenuated Wnt/β-catenin signaling and increased expression of Wnt signaling antagonists including Dkk-1 and sFRP-1. These studies strongly suggest that glucocorticoid use causes deregulation of Wnt/β-catenin signaling, consequently resulting in the associated bone loss and marrow adiposity.

IV. Wnt/β-Catenin Signaling As a Therapeutic Target?

In order to alleviate the bone loss and associated deregulation of cell populations of the bone marrow following damaging treatments as described above, the development of specifically targeted therapies is required. Considering its widespread involvement and important roles in increasing bone formation and suppressing fat formation, the Wnt/β-catenin signaling pathway presents a potential target for future therapies to treat bone loss conditions such as ageing/menopause-induced osteoporosis, damaging therapy-induced bone defects and other such ailments. However, as Wnt/β-catenin signaling is a regulatory pathway involved in proliferation and differentiation, potential over-activation leading to tumorigenesis is potentially of concern. In order to address this, there have been a number of investigations into targeting the secreted antagonists of the pathway, particularly Dkk-1, sclerostin and to a lesser extents FRP-1. In mouse models, these antagonists are primarily expressed by osteoblasts or osteocytes (Poole, van Bezooijen et al., 2005; Li, Sarosi et al., 2006; Baron and Rawadi 2007), thus systemic administration of targeted treatments would not disrupt Wnt signaling in other organs/cell types, making them attractive therapeutic targets.

Sclerostosis, which is due to the insufficiency of sclerostin production, is a disease characterized by high bone mass (Hamersma, Gardner et al., 2003). Utilizing this knowledge, knockout of the sclerostin gene, SOST, in animal studies have result in high bone mass, which has provided the basis for further investigations into potential therapeutic applications (Li, Ominsky et al., 2008; Ke, Richards et al., 2012). Inhibition of sclerostin using a specific monoclonal antibody (Scl-Ab) in animal studies has demonstrated its potential in limiting bone loss and/or increasing bone formation, preventing the associated bone loss caused by estrogen deficiency in an ovariectomy model, as well as increasing bone mass and serum osteocalcin in male rats (Li, Ominsky et al., 2009; Li, Warmington et al., 2010). Furthermore, in a non-human primate model, systemic administration of the sclerostin antibody for 10 weeks resulted in increases in BMD and strength throughout the entire skeleton, including sites known to be susceptible to osteoporosis, such as the lumbar spine, total hip and distal radius (Ominsky, Li et al., 2011). The first clinical study using a humanized sclerostin monoclonal antibody known as AMG785 was performed on healthy men and postmenopausal women, administering a single IV or SC dose at varying concentrations (Padhi, Jang et al., 2011). Dose-dependent increases in bone formation markers propeptide of type 1 collagen, bone ALP and osteocalcin were observed, as well as an increase in BMD of the lumbar spine and total hip in all cohorts. Interestingly the effect of a single dose of the sclerostin antibody was found to be comparable to that of 6 months of daily parathyroid hormone (PTH) treatment (Padhi, Jang et al., 2011). The above studies have demonstrated that an anti-sclerostin antibody represents a promising therapy for postmenopausal and ageing-induced osteoporosis and other bone loss conditions. However, further investigations into the effects of Scl-Ab treatment on other cell types/tissues and in longer-term treatment models are required to examine any potential chronic side effects.

In addition to the above studies investigating sclerostin as a potential target, Wnt antagonists Dkk-1 and sFRP-1 have also been pursued. *In vitro* and *in vivo* investigations have observed a Dkk-1 neutralizing antibody to increase osteogenic differentiation of a mesenchymal-like cell line (CH10T1/2) and increase trabecular bone mineral density (BMD) in growing female mice (Glantschnig, Hampton et al., 2010). In a follow-up study, 8 weeks of treatment with this anti-Dkk-1 antibody alleviated OVX-induced bone loss revealed by increased femoral and lumbar spine BMD, demonstrating its potential use in the treatment of postmenopausal osteopenia or osteoporosis (Glantschnig, Scott et al., 2011). Dkk-1 has also been extensively studied in the setting of multiple myeloma-induced bone disease and bone loss conditions and recently a novel anti-Dkk-1 neutralising monoclonal antibody was examined for its effect on multiple myeloma (MM) and the bone microenvironment (Pozzi, Fulciniti et al., 2013). The use of the Dkk-1 antibody not only prevented MM-induced osteolysis in a mouse model but increased osteoblast numbers on bone surfaces, demonstrating Dkk-1 as a potentially effective therapeutic target (Pozzi, Fulciniti et al., 2013).

Investigations into sFRP-1 inhibition have also been successful. For example, an orally bioavailable small molecule inhibitor of sFRP-1 has been developed. It disrupts the protein-protein interaction between Wnts and sFRP-1, thus allowing more Wnt ligands to bind the LRP5/6 receptor to increase bone formation (Moore, Kern et al., 2009). *Ex vivo* mouse calvarial cultures with this compound exhibited increased osteoblast numbers and areas of new bone formation when examined histologically (Bodine, Stauffer et al., 2009;

Moore, Kern et al., 2009). Although remaining in the early stages, studies such as these illustrate a promising future for therapies targeting the Wnt/β-catenin signaling pathway to promote bone formation and in turn enable appropriate regulation of the bone marrow microenvironment. Although a recent study has suggested that preservation of Wnt/β-catenin signaling can maintain bone formation and prevent bone marrow adiposity (Georgiou, King et al., 2012) during methotrexate chemotherapy in rats, further studies are required to investigate whether the Wnt/β-catenin signaling pathway can be a therapeutic target for preventing cancer chemotherapy-induced bone loss and bone marrow adiposity.

Conclusion

Conditions of stress or injury induced pharmacologically, caused by cancer chemotherapy and glucocorticoid therapy in particular, alter steady-state cellular composition of the bone marrow microenvironment and in turn its function. This results in deregulation of stromal cell commitment and differentiation, favoring adipogenesis at the expense of osteogenesis. While the mechanisms underlying the significant increase in marrow fat and bone loss remain to be clearly identified, it has been established that there is a network of signaling pathways that act in conjunction to enable an appropriately functioning bone marrow. The Wnt/β-catenin signaling pathway not only promotes bone formation, but inhibits adipogenic differentiation. Wnt10b has been extensively investigated for its role in increasing osteogenesis by inducing Runx2 while inhibiting adipogenesis through PPAR-γ and C/EBPα. Wnt antagonists, of either the Wnt ligands themselves or of the LRP5/6 co-receptors, including sclerostin, Dkk-1, sFRP-1 and Wif-1, have been illustrated to play a role in regulating the lineage commitment and differentiation of bone marrow stromal progenitor cells, inducing preferential adipogenic commitment and differentiation.

As described herein, deregulation of components of the Wnt/β-catenin signaling pathway contributes to the bone loss and marrow adiposity observed under damaging conditions such as chemotherapy, glucocorticoid use and with ageing. As a result of its widespread involvement, the Wnt/β-catenin signaling pathway may be a suitable potential target for promoting bone formation and reducing fat formation. Although it remains to be investigated whether this signaling pathway may be explored as a potential target for preserving bone formation and preventing marrow adiposity during cancer chemotherapy and glucocorticoid use, recent preclinical investigations into the use of a neutralizing antibody against Wnt antagonist sclerostin for the treatment of osteoporosis have established an effect on alleviating estrogen-deficiency-induced bone loss.

Despite the investigations reviewed herein, future studies are still required to gain a better understanding of the deregulation caused to the Wnt/β-catenin pathway in order to assess its specific roles in the damage and recovery of bone marrow cell populations. In order to refine current treatments and develop more specifically targeted therapies to reestablish a defective bone marrow as well as prevent bone loss, a clearer understanding is required of the complex mechanisms that regulate commitment and differentiation of bone marrow cell populations.

Acknowledgments

This book chapter has reviewed some of the authors' own work funded in part by Bone Health Foundation (Australia), Channel-7 Children's Research Foundation of South Australia, and National Health and Medical Research Council (NHMRC) of Australia.

References

Abdallah, B. M. and M. Kassem (2012). "New factors controlling the balance between osteoblastogenesis and adipogenesis." *Bone* 50(2): 540-545.

Anastasilakis, A. D., S. A. Polyzos, et al., (2010). "The effect of teriparatide on serum Dickkopf-1 levels in postmenopausal women with established osteoporosis." *Clinical endocrinology* 72(6): 752-757.

Baron, R. and G. Rawadi (2007). "Targeting the Wnt/beta-catenin pathway to regulate bone formation in the adult skeleton." *Endocrinology* 148(6): 2635-2643.

Belaid-Choucair, Z., Y. Lepelletier, et al., (2008). "Human bone marrow adipocytes block granulopoiesis through neuropilin-1-induced granulocyte colony-stimulating factor inhibition." *Stem Cells* 26(6): 1556-1564.

Bennett, C. N., K. A. Longo, et al., (2005). "Regulation of osteoblastogenesis and bone mass by Wnt10b." *Proc. Natl. Acad. Sci. USA* 102(9): 3324-3329.

Bianco, P. (2011). "Minireview: The stem cell next door: skeletal and hematopoietic stem cell "niches" in bone." *Endocrinology* 152(8): 2957-2962.

Bodine, P. V., J. Billiard, et al., (2005). "The Wnt antagonist secreted frizzled-related protein-1 controls osteoblast and osteocyte apoptosis." *Journal of cellular biochemistry* 96(6): 1212-1230.

Bodine, P. V., B. Stauffer, et al., (2009). "A small molecule inhibitor of the Wnt antagonist secreted frizzled-related protein-1 stimulates bone formation." *Bone* 44(6): 1063-1068.

Bonewald, L. F. and M. L. Johnson (2008). "Osteocytes, mechanosensing and Wnt signaling." *Bone* 42(4): 606-615.

Brunetti, G., M. F. Faienza, et al., (2013). "High dickkopf-1 levels in sera and leukocytes from children with 21-hydroxylase deficiency on chronic glucocorticoid treatment." *American journal of physiology - Endocrinology and metabolism* 304(5): E546-554.

Butler, J. S., J. M. Queally, et al., (2010). "Silencing Dkk1 expression rescues dexamethasone-induced suppression of primary human osteoblast differentiation." *BMC musculoskeletal disorders* 11: 210.

Buttiglieri, S., M. Ruella, et al., (2011). "The aging effect of chemotherapy on cultured human mesenchymal stem cells." *Exp. Hematol.* 39(12): 1171-1181.

Cawthorn, W. P., A. J. Bree, et al., (2012). "Wnt6, Wnt10a and Wnt10b inhibit adipogenesis and stimulate osteoblastogenesis through a beta-catenin-dependent mechanism." *Bone* 50(2): 477-489.

Chaudhary, L., H. A., et al., (2004). "Differential growth factor control of bone formation through osteoprogenitor differentiation." *Bone* 34(3): 402-411.

Chung, S. S., J. S. Lee, et al., (2012). "Regulation of Wnt/beta-catenin signaling by CCAAT/enhancer binding protein beta during adipogenesis." *Obesity* 20(3): 482-487.

Despars, G. and Y. St-Pierre (2011). "Bidirectional interactions between bone metabolism and hematopoiesis." *Experimental hematology* 39 (8): 809-816.

Fan, C., J. C. Cool, et al., (2009). "Damaging effects of chronic low-dose methotrexate usage on primary bone formation in young rats and potential protective effects of folinic acid supplementary treatment." *Bone* 44(1): 61-70.

Fan, C., K. R. Georgiou, et al., (2011). "Methotrexate toxicity in growing long bones of young rats: a model for studying cancer chemotherapy-induced bone growth defects in children." *Journal of biomedicine & biotechnology* 2011: 903097.

Fogelman, I., G. M. Blake, et al., (2003). "Bone mineral density in premenopausal women treated for node-positive early breast cancer with 2 years of goserelin or 6 months of cyclophosphamide, methotrexate and 5-fluorouracil (CMF)." *Osteoporosis international* 14(12): 1001-1006.

Georgiou, K. R., B. K. Foster, et al., (2010). "Damage and recovery of the bone marrow microenvironment induced by cancer chemotherapy - potential regulatory role of chemokine CXCL12/receptor CXCR4 signalling." *Curr. Mol. Med.* 10(5): 440-453.

Georgiou, K. R., S. K. Hui, et al., (2012). "Regulatory pathways associated with bone loss and bone marrow adiposity caused by aging, chemotherapy, glucocorticoid therapy and radiotherapy." *Am. J. Stem.* Cell. 1(3): 205-224.

Georgiou, K. R., T. J. King, et al., (2012). "Attenuated Wnt/beta-catenin signalling mediates methotrexate chemotherapy-induced bone loss and marrow adiposity in rats." *Bone* 50(6): 1223-1233.

Georgiou, K. R., M. A. Scherer, et al., (2012). "Methotrexate chemotherapy reduces osteogenesis but increases adipogenic potential in the bone marrow." *Journal of cellular physiology* 227(3): 909-918.

Glantschnig, H., R. A. Hampton, et al., (2010). "Generation and selection of novel fully human monoclonal antibodies that neutralize Dickkopf-1 (DKK1) inhibitory function in vitro and increase bone mass in vivo." *The Journal of biological chemistry* 285(51): 40135-40147.

Glantschnig, H., K. Scott, et al., (2011). "A rate-limiting role for Dickkopf-1 in bone formation and the remediation of bone loss in mouse and primate models of postmenopausal osteoporosis by an experimental therapeutic antibody." *The Journal of pharmacology and experimental therapeutics* 338(2): 568-578.

Hamersma, H., J. Gardner, et al., (2003). "The natural history of sclerosteosis." *Clinical genetics* 63(3): 192-197.

He, J., T. Sheng, et al., (2006). "Suppressing Wnt signaling by the hedgehog pathway through sFRP-1." *The Journal of Biological Chemistry* 281(47): 35598-35602.

Hung, S. H., C. H. Yeh, et al., (2008). "Pioglitazone and dexamethasone induce adipogenesis in D1 bone marrow stromal cell line, but not through the peroxisome proliferator-activated receptor-gamma pathway." *Life sciences* 82(11-12): 561-569.

Ke, H. Z., W. G. Richards, et al., (2012). "Sclerostin and Dickkopf-1 as therapeutic targets in bone diseases." *Endocrine reviews* 33(5): 747-783.

King, T. J., K. R. Georgiou, et al., (2012). "Methotrexate Chemotherapy Promotes Osteoclast Formation in the Long Bone of Rats via Increased Pro-Inflammatory Cytokines and Enhanced NF-kappaB Activation." *The American journal of pathology.*

Kitajima, M., M. Shigematsu, et al., (2007). "Effects of glucocorticoid on adipocyte size in human bone marrow." *Medical molecular morphology* 40(3): 150-156.

Kitazawa, R., K. Mori, et al., (2008). "Modulation of mouse RANKL gene expression by Runx2 and vitamin D3." *J. Cell. Biochem.* 105(5): 1289-1297.

Li, H. X., X. Luo, et al., (2008). "Roles of Wnt/beta-catenin signaling in adipogenic differentiation potential of adipose-derived mesenchymal stem cells." *Mol. Cell. Endocrinol.* 291(1-2): 116-124.

Li, J., I. Sarosi, et al., (2006). "Dkk1-mediated inhibition of Wnt signaling in bone results in osteopenia." *Bone* 39(4): 754-766.

Li, X., M. S. Ominsky, et al., (2008). "Targeted deletion of the sclerostin gene in mice results in increased bone formation and bone strength." *Journal of bone and mineral research* 23(6): 860-869.

Li, X., M. S. Ominsky, et al., (2009). "Sclerostin antibody treatment increases bone formation, bone mass, and bone strength in a rat model of postmenopausal osteoporosis." *Journal of bone and mineral research* 24(4): 578-588.

Li, X., K. S. Warmington, et al., (2010). "Inhibition of sclerostin by monoclonal antibody increases bone formation, bone mass, and bone strength in aged male rats." *Journal of bone and mineral research* 25(12): 2647-2656.

Lin, L., S. D. Dai, et al., (2010). "Glucocorticoid-induced differentiation of primary cultured bone marrow mesenchymal cells into adipocytes is antagonized by exogenous Runx2." *Acta pathologica, microbiologica, et immunologica Scandinavica* 118(8): 595-605.

Long, M. (2001). "Osteogenesis and Bone-Marrow-Derived cells." *Blood cells, molecules and diseases* 27(3): 677-690.

Macsai, C. E., B. K. Foster, et al., (2008). "Roles of Wnt signalling in bone growth, remodelling, skeletal disorders and fracture repair." *J. Cell. Physiol.* 215(3): 578-587.

Michal, S. A., D. J. Adelstein, et al., (2012). "Multi-agent concurrent chemoradiotherapy for locally advanced head and neck squamous cell cancer in the elderly." *Head & neck* 34(8): 1147-1152.

Moerman, E. J., K. Teng, et al., (2004). "Aging activates adipogenic and suppresses osteogenic programs in mesenchymal marrow stroma/stem cells: the role of PPAR-gamma2 transcription factor and TGF-beta/BMP signaling pathways." *Aging cell* 3(6): 379-389.

Moore, W. J., J. C. Kern, et al., (2009). "Modulation of Wnt signaling through inhibition of secreted frizzled-related protein I (sFRP-1) with N-substituted piperidinyl diphenylsulfonyl sulfonamides." *Journal of medicinal chemistry* 52(1): 105-116.

Muruganandan, S., A. A. Roman, et al., (2009). "Adipocyte differentiation of bone marrow-derived mesenchymal stem cells: cross talk with the osteoblastogenic program." *Cellular and molecular life sciences* 66(2): 236-253.

Naito, M., K. Omoteyama, et al., (2012). "Inhibition of Wnt/beta-catenin signaling by dexamethasone promotes adipocyte differentiation in mesenchymal progenitor cells, ROB-C26." *Histochemistry and cell biology* 138(6): 833-845.

Naveiras, O., V. Nardi, et al., (2009). "Bone-marrow adipocytes as negative regulators of the haematopoietic microenvironment." *Nature* 460(7252): 259-263.

Ominsky, M. S., C. Li, et al., (2011). "Inhibition of sclerostin by monoclonal antibody enhances bone healing and improves bone density and strength of nonfractured bones." *Journal of bone and mineral research* 26(5): 1012-1021.

Padhi, D., G. Jang, et al., (2011). "Single-dose, placebo-controlled, randomized study of AMG 785, a sclerostin monoclonal antibody." *Journal of bone and mineral research* 26(1): 19-26.

Peled, A., I. Petit, et al., (1999). "Dependence of human stem cell engraftment and repopulation of NOD/SCID mice on CXCR4." *Science* 283: 845-848.

Poncin, G., A. Beaulieu, et al., (2012). "Characterization of spontaneous bone marrow recovery after sublethal total body irradiation: importance of the osteoblastic/adipocytic balance." *PloS one* 7(2): e30818.

Poole, K. E., R. L. van Bezooijen, et al., (2005). "Sclerostin is a delayed secreted product of osteocytes that inhibits bone formation." *FASEB journal* 19(13): 1842-1844.

Pozzi, S., M. Fulciniti, et al., (2013). "In vivo and in vitro effects of a novel anti-Dkk1 neutralizing antibody in multiple myeloma." *Bone* 53(2): 487-496.

Raghu Nadhanan, R., S. M. Abimosleh, et al., (2012). "Dietary emu oil supplementation suppresses 5-fluorouracil chemotherapy-induced inflammation, osteoclast formation, and bone loss." *American journal of physiology - Endocrinology and metabolism* 302(11): E1440-1449.

Ross, S. E., N. Hemati, et al., (2000). "Inhibition of adipogenesis by Wnt signaling." *Science* 289(5481): 950-953.

Saag, K. G., E. Shane, et al., (2007). "Teriparatide or alendronate in glucocorticoid-induced osteoporosis." *The New England journal of medicine* 357(20): 2028-2039.

Scheideler, M., C. Elabd, et al., (2008). "Comparative transcriptomics of human multipotent stem cells during adipogenesis and osteoblastogenesis." *BMC Genomics* 9: 340.

Stevens, S. K., S. G. Moore, et al., (1990). "Early and late bone-marrow changes after irradiation: MR evaluation." *Am. J. Roentgenol.* 154(4): 745-750.

Taipaleenmaki, H., B. M. Abdallah, et al., (2011). "Wnt signalling mediates the cross-talk between bone marrow derived pre-adipocytic and pre-osteoblastic cell populations." *Exp. Cell. Res.* 317(6): 745-756.

Wang, F. S., J. Y. Ko, et al., (2008). "Modulation of Dickkopf-1 attenuates glucocorticoid induction of osteoblast apoptosis, adipocytic differentiation, and bone mass loss." *Endocrinology* 149(4): 1793-1801.

Xian, C., J. Cool, et al., (2006). "Damage and recovery of the bone growth mechanism in young rats following 5-fluorouracil acute chemotherapy." *Journal Cellular Biochemistry* 99(6): 1688-1704.

Xian, C. J., J. C. Cool, et al., (2008). "Folinic acid attenuates methotrexate chemotherapy-induced damages on bone growth mechanisms and pools of bone marrow stromal cells." *J. Cell. Physiol.* 214(3): 777-785.

Yao, W., Z. Cheng, et al., (2008). "Glucocorticoid excess in mice results in early activation of osteoclastogenesis and adipogenesis and prolonged suppression of osteogenesis: a longitudinal study of gene expression in bone tissue from glucocorticoid-treated mice." *Arthritis and rheumatism* 58(6): 1674-1686.

Zhou, H., W. Mak, et al., (2008). "Osteoblasts directly control lineage commitment of mesenchymal progenitor cells through Wnt signaling." *J. Biol. Chem.* 283(4): 1936-1945.

In: Adipogenesis
Editors: Y. Lin and X. Cai

ISBN: 978-1-62808-750-5

Chapter V

Role of Reactive Oxygen Species in Adipocyte Differentiation

***D. Lettieri-Barbato*[1], *K. Aquilano*[1], *G. Tatulli*[3] *and M. R. Ciriolo*[1,2]**

[1]Department of Biology, University of Rome Tor Vergata, Rome, Italy
[2]IRCCS San Raffaele "La Pisana", Rome, Italy
[3]Università Telematica San Raffaele Rome, Italy

Abstract

Compelling evidence demonstrates a relationship between reactive oxygen species (ROS) production and adipocytes differentiation; however, no clear proofs about the genuine source(s) of ROS during adipogenesis are available. The synchronized initiation of adipogenesis and mitochondrial biogenesis indicates that mitochondria play a pertinent role in the differentiation and maturation of adipocytes. The early stages of mitochondrial biogenesis and adipocytes differentiation are strongly related to enhanced ROS production. On the basis of this evidence, it is likely that mitochondrial-derived ROS could be mainly involved in the initiation of the redox cascade triggering adipocytes differentiation. Intriguingly, ROS are essential to activate the transcriptional machinery necessary to evoke adipogenesis. Here we discuss how adequate levels of ROS maintain cellular homeostasis by creating a suitable redox environment that allows and sustains pre-adipocyte differentiation without causing cellular oxidative damage.

Introduction

Adipose tissue (AT) represents the larger tissue in body. AT responds rapidly and dynamically to alteration in nutrient levels remodelling its mass [1]. These adaptations confer to AT the capacity to function as the main energy sink in the body.

Using incorporation of environmental ^{14}C as a tracer, Spalding and co-workers have recognized that "new adipocytes form constantly to replace lost adipocytes" and estimated the average of half-life of the adipocyte to be in the order of 8.3 years [2]. This research group postulated that adipocyte number is fairly fixed by early adulthood, and that alterations in fat mass during adulthood are merely credited to alterations in adipocyte size [3]. The percentage of newly emerging adipocytes is balanced by adipocyte death, with the total number of adipocytes being tightly controlled, suggesting that whole system is in a state of constant flux. This suggests that pre-adipocytes or adipocyte progenitor cells are recruited into stromal vascular fraction of AT and dynamically propagate into mature adipocytes (postmitotic cells) by an adipogenesis process [4]. Adipogenesis is characterized by an orchestrated cascade of molecular events that directly activate transcription of genes involved in pre-adipocytes differentiation. Although several works have elucidated the transcriptional program that induces adipogenesis, the signalling mechanisms underlying the activation of transcriptional machinery are not fully understood.

1. Adipogenesis: A Dynamic Process Governed by ROS

Adipogenesis is a process commonly divided in: *i*) an early phase, in which confluent adipocytes precursors undergo a mitotic clonal expansion (MCE) doubling their numbers; *ii*) a late phase in which pre-adipocytes transform into lipid-laden mature adipocytes. Among cellular mediators involved in adipogenesis, ROS seem to play a leading role. It has been demonstrated that in human adipose-derived stem pre-adipocytes as well as in murine pre-adipocytes, adipogenic differentiation is accompanied by increased ROS production. Coherently, scavenging of ROS production inhibits adipogenesis [5]. It has been suggested that a transient burst of low amounts of ROS, or a chronic mild pro-oxidant conditions, is generally related to differentiation of pre-adipocytes in lipid-laden mature adipocytes. Relatively to this aspect, several works showed that the creation of an adequate pro-oxidant milieu in pre-adipocytes facilitates their differentiation [6, 7]. Glutathione (GSH) is considered the most important non-enzymatic antioxidant compound that serves for modulating the redox status of protein thiols, detoxification and direct scavenging activity of oxyradicals [8]. GSH can be oxidized (GSSG) when cells are exposed to increased levels of oxidative stress [9]. Recently, we have revealed that, during adipogenesis of murine 3T3-L1 cells, the GSH/GSSG ratio decreases, shifting redox status towards oxidizing conditions. In order to clarify the role of GSH in adipogenesis, we administered buthionine sulfoximine (BSO), a well-accepted inhibitor of GSH synthesis. Interestingly, BSO treatment improves the early phase of adipogenesis and induces higher triglyceride accumulation in differentiated adipocytes [9]. Contrarily, GSH ethyl ester (GSHest) supplementation abrogates this process also in the presence of BSO [9]. Similarly, chronic administration of hydrogen peroxide (H_2O_2) also stimulates the initial phase of adipogensis [6]. During these phases, a coordinate roundup of redox sensitive transcription factors build well differentiated insulin sensitive adipocytes. Three members of the CAAT/enhancer binding protein (C/EBP)1 family, C/EBPβ, C/EBPα, and CHOP-10/GADD153 (C/EBP homologous protein also identified as growth arrest and DNA damage 153) are early expressed in the differentiating adipocytes. During the MCE phase, CHOP-10/GADD153 expression falls and, simultaneously, C/EBPβ and C/EBPα are transiently induced. This event mediates the later expression of C/EBPα and peroxisome proliferation-activated receptor-γ (PPARγ), subsequently triggering full-blown adipocyte differentiation [9]. It has been suggested that DNA binding activity of C/EBPβ can be positively influenced by its double phosphorylation by mitogen-activated protein kinases (MAPK) and glycogen synthase kinase 3 beta (GSK3β), which are kinases notably regulated also by a redox-dependent mechanism [6]. In particular, Thr188 of C/EBPβ is phosphorylated by MAPK immediately after its expression is induced and then later by GSK3β on Ser184 or Thr179. Mild oxidizing conditions promptly trigger C/EBPβ phosphorylation enhancing its binding activity on PPARγ gene promoter. This suggests that the phosphorylation of C/EBPβ leads to a conformational change that facilitates dimerization, letting the basic DNA-binding region of C/EBPβ accessible to the C/EBP regulatory element on C/EBPα or PPARγ. Furthermore, mild oxidizing conditions facilitate S–S bond formation and dimerization of C/EBPβ, letting the basic region accessible to the C/EBP regulatory element [10]. In relation to this aspect, the thioredoxin-interacting protein (TXNIP) has been recently demonstrated to influence adipocyte development *in vivo*. Specifically, loss of TXNIP, which inhibits the antioxidant protein thioredoxin involved in reducing thiol groups, increases adipogenesis in culture and adiposity *in vivo* [11]. This leads to improved insulin sensitivity through increased PPARγ expression and activity. Thus, the oxidative status that we have demonstrated occurring during normal differentiation as well as by BSO treatment, likely creates an oxidative modification of C/EBPα, thus facilitating the expression of its downstream adipogenic genes (Figure 1A).

Relatively to redox modulation of adipogenesis, a plethora of studies have suggested the use of some phytochemicals as potential modulators of adipocyte differentiation. It has been hypothesised that this function is mediated by the antioxidant potential of some plant-derived molecules [12]. In particular, resveratrol (3,4',5-trihydroxystilbene) represents the most studied natural compound in the field of adipogenesis modulation. Resveratrol is found in various plants, including grapes, berries and peanuts. It is also present in wine, especially red wine. During the last years, numerous *in vitro* and *in vivo* studies have revealed that resveratrol has an anti-adipogenic effect that mainly derives from its antioxidant properties [9]. It has been suggested that low doses of resveratrol efficiently prevent MCE and terminal adipogenesis in 3T3-L1 pre-adipocytes. This function can be attributed to its ability in strengthening intracellular antioxidant defence mainly by increasing GSH content through γ-glutamyl-cysteine ligase (γGCL) induction [9]. Moreover, resveratrol inhibits AKT and MAPK signalling, thus down-regulating cyclin D1 expression, and subsequently inhibiting cell cycle re-entry and MCE [13, 14]. Similarly, also resveratrol mimetic compounds, such as RSVA314 and RSVA405, inhibit 3T3-L1 adipocytes differentiation by interfering with MCE. In particular, RSVA314 and RSVA405 prevent the transcriptional changes of multiple gene products involved in the adipogenic process, including PPARγ and C/EBPα. Furthermore, orally administered RSVA405 significantly reduce the body weight gain of mice fed with a high-fat diet [15]. Recently we have reported that resveratrol also modulate antioxidant defence by increasing GSH levels in pre-adipocytes, thus indirectly counteracting adipogenesis [9]. Interestingly, it has been suggested that resveratrol also attenuates high-fat diet (HFD)-induced adipogenesis in the epididymal fat tissues of mice [16]. In comparison with HFD-fed mice, mice fed with a 0.4% resveratrol-supplemented diet (RSD) show significant lower body weight gain (−48%), visceral fat-pad weights (−58%), plasma levels of triglyceride, free-fatty acids (FFA), total cholesterol and glucose. Resveratrol significantly reverse the HFD-induced up-regulation of adipogenic genes (PPARγ and C/EBPα) in the epididymal adipose tissue [16]. However, independently of its antioxidant capability, resveratrol modulates adipogenesis directly orchestrating some molecular axis involving both nuclear NAD^+-dependent protein deacetylase sirtuin 1 (Sirt1) and AMP-activated protein kinase (AMPK). Sirt1 plays important roles in a wide variety of processes, including stress resistance, energy metabolism and differentiation [17, 18]. Sirt1 indirectly influences the transcriptional activity of the nuclear receptor PPARγ by docking the nuclear receptor co-repressors NCoR and SMRT to PPARγ. This mechanism was shown to be responsible for the ability of Sirt1 to mobilize fat in fully differentiated adipocytes and also attenuate development of adipocytes from pre-adipocytes [19]. Recently, Carl-Magnus Bäckesjö and co-workers have found that resveratrol-mediated activation of Sirt1 in mesenchimal stem cells (C3H10T1/2) decreases adipocyte differentiation by PPARγ inhibition [20]. Similarly to Sirt1, also AMPK is a key modulator of adipocyte differentiating process. AMPK is known to sense the cellular energy state and regulates various cellular energy metabolism pathways through its activation by AMP, an indicator of a low-energy state [21]. Incubation of 3T3-L1 pre-adipocytes with resveratrol enhances phospho-activation of AMPK in a dose-dependent manner, while total AMPK levels were unchanged, and PPARγ and C/EBPα levels were decreased. Interestingly, pre-treatment with AMPK RNA interference (siRNA) and resveratrol promotes PPARγ and C/EBPα adipocyte differentiation, while the decrease of p-AMPK increases protein expression [22]. Furthermore, also resveratrol mimetic compounds such as RSVA314 and RSVA405 act as potent indirect activators of AMPK. Indeed, similarly to resveratrol, RSVA314 and RSVA405 significantly activate AMPK and inhibit acetyl-CoA carboxylase (ACC), one target of AMPK and a key regulator of fatty acid biogenesis in proliferating 3T3-L1 adipocytes. Also intraperitoneal administration of AICAR, a selective agonist of AMPK, to mice fed with HFD (60% kcal% fat) significantly blocks the body weight gain and total content of epididymal fat over a period of 6 weeks [23]. AICAR has the ability to inhibit adipocyte differentiation mainly enhancing β-catenin expression and its nuclear accumulation. The expression of the major genes of adipogenesis including the PPARγ and C/EPBα, which were all reduced following AICAR treatment, is significantly recovered in β-catenin siRNA-transfected cells [24].

1.1. Potential Mechanisms by Which Ros Enhance Adipogenesis

It is well known that adipocytes represent the main insulin sensitive cells in the body and insulin represents a key hormone triggering adipogenic start-up [25-27]. The involvement of an oxidation step in the insulin pathway has been suggested for decades, but only recently insulin-induced ROS have been identified as key messengers facilitating the insulin-signalling cascade. Insulin itself was shown to elicit the generation of H_2O_2 in adipocytes by activating a plasma membrane enzyme system with the properties of a NADPH oxidase (Nox) [28, 29]. Nox is a multimeric enzyme, and seven isoforms of the catalytic subunit have been identified in mammalian cells. All isoforms have the catalytic domain that allows transport of electron from cytosolic NADPH to generate superoxide ($O_2^{\bullet-}$), which is rapidly converted in H_2O_2 by superoxide dismutase (SOD). Among all Nox isoforms, Nox4 is highly expressed in pre-adipocytes [30] and is involved in modulating pre-adipocyte proliferation and/or differentiation [7, 31]. Actually, *pan*-Nox inhibitors efficiently block adipogenesis [31]. Several works have focused their attention on the Nox4 isoform. In particular Kanda et al., found that in mesenchymal stem cells, Nox4-produced intracellular ROS enhance adipogenic differentiation [7]. Similarly, Schröder and co-workers also showed that inhibiting Nox4 expression blocks pre-adipocytes differentiation [31]. These findings propose that Nox4 may have a positive role in promoting adipogenesis, probably by facilitating insulin signalling [29]. Unlike other Nox proteins, Nox4 can generate ROS under basal conditions in the absence of exogenous insulin stimuli [32]. Interestingly, Nox4 is the only isoform that primarily produces H_2O_2 instead of $O_2^{\bullet-}$ [32], and this property may be attributable to an extracellular domain called E-loop, which is longer than that of Nox1 or Nox2 [33].

Several proteins sensitive to redox regulation participate in insulin cascade [34]. The presence of redox-sensitive sulfhydryl groups in the key cysteine residues in these signalling proteins or their binding partners is essential for their responsiveness to insulin-mediated ROS production. Redox modulation of sulfhydryl groups may act as "on-off" switch for changing structural configuration of the protein and thereby affecting the activity/affinity of the target. Principal targets of insulin-mediated ROS are the protein tyrosine phosphatases (PTPs) [34-36]. The PTP super family includes phosphatases such as the prototypic PTP1B that dephosphorylates tyrosyl phosphorylated substrates and enzymes such as PTEN, which dephosphorylates phosphatidylinositol-3,4,5-triphosphate (PIP3) to terminate PI3K signalling [37]. The characteristic of the active site of PTPs makes them highly susceptible to reversible oxidation by H_2O_2 [36]. The oxidation of the active site containing cysteine abrogates its nucleophilic properties, thus rendering PTPs inactive. PTPs such as PTP1B and PTEN that modulate insulin sensitivity *in vivo* [38, 39] are transiently oxidized by H_2O_2 in response to insulin in cell culture systems and promote insulin-receptor activation and PI3K signalling cascade, respectively [29, 40-42]. Upon insulin stimulation, Nox-derived ROS are key modulators of insulin signalling cascade via the oxidative inhibition of cellular PTP activity. Indeed, differentiated 3T3-L1 adipocytes display a high H_2O_2 production following insulin stimulation [43, 44]. In 3T3-L1 adipocytes, antioxidant enzyme catalase as well as dithiothreitol (DTT) block the oxidation of PTP induced by insulin, suggesting that the oxidant signal inhibits cellular PTPs that serve as negative regulators of the insulin signalling cascade [43]. Accordingly, also diphenyl-iodonium (DPI), an inhibitor of cellular Nox, dampens the insulin-mediated activation both of PI 3-kinase activity and serine kinase Akt [44]. The H_2O_2-induced activation of Akt is mediated by upstream activation of PI 3′-kinase activity, because treatment of 3T3-L1 adipocytes with the PI 3′-kinase inhibitors (wortmannin or LY294002) completely blocks the activation of Akt by exogenous H_2O_2. These data suggest that cellular PI 3′-kinase is a critical upstream mediator of Akt activation by oxidative molecules in a variety of signalling pathways, including insulin action (Figure 1B).

Although tremendously high levels of ROS unquestionably cause cellular damage [45], an adequate ROS production ensure cellular homeostasis that permits and sustains pre-adipocyte differentiation without inflicting cellular damage.

2. The Central Role of Mitochondria in Adipogenesis

The adipocyte displays a distinctive structure generally recognizable by a large lipid droplet associated with relatively low cytoplasmic volume and reduced mitochondrial density. Despite adipocytes contain relatively low mitochondrial mass compared to its overall size, their mitochondria play an essential role in many different pathways. Indeed, adipocyte mitochondria interpret the nutrient levels coordinating their responses oxidizing FAs and carbohydrates fuels through the tricarboxylic acid (TCA) cycle and the respiratory chain. Alternatively, adipocyte stores these fuels safely in the form of triglycerides until whole-body energy requirements signal for their release [46]. Interestingly, the synchronized initiation of adipogenesis and mitochondrial biogenesis indicates that mitochondria play also a relevant role in the differentiation of adipocytes [47]. A recent study by Tormos and colleagues confirms that the early events of enhanced mitochondrial metabolism, biogenesis and ROS production are crucial for the initiation and promotion of adipocyte differentiation [48]. In particular, these authors have reported that low levels of mitochondrial-generated ROS are essential to regulate adipogenesis. During adipocyte differentiation there is mTORC1-dependent increase in mitochondrial metabolism and biogenesis. A consequence of this increase is the production of mitochondrial complex III-derived ROS, which are transduced in the activation of PPARγ transcriptional machinery required to promote adipocyte differentiation.

2.1. Ros As Direct Products of Mitochondrial Metabolism

In a physiological context, mitochondrial ROS are generally produced at a low level by the electron transport chain (ETC) as a normal part of cellular metabolism [46, 49-50]. However, ROS production can rise, with changes in oxidative mitochondrial metabolism, such as during nutrient excess [51, 52]. Nutrient flux targets TCA and resulting electrons are transferred to nicotinamide adenine dinucleotide (NAD^+) and flavin adenine dinucleotide (FAD) to produce NADH and $FADH_2$. These electrons are donated to ETC at complexes I and II, respectively. Electrons are shuttled from complexes I and II to complex III via ubiquinone, and from complex III to complex IV via cytochrome *c*. Two electrons are finally donated to molecular oxygen (1/2 O_2), generating H_2O. Incomplete, one-electron reduction of oxygen can also occur at complexes I, II, and III, producing $O_2^{\bullet -}$.When breakdown of metabolites in the TCA cycle exceeds the capacity of the ETC to assimilate the resulting electrons, ROS production increases. While $O_2^{\bullet -}$ mediates its effects within a short range of its production, it can also be converted by SOD into H_2O_2 that is more stable and can diffuse throughout the cell.

Recently, the generally accepted idea that complexes of the ETC are the 'major' sites of ROS production in the mitochondria has been challenged. Independent studies have demonstrated that also α-ketoglutarate dehydrogenase (α–KGDH) could be an important source of ROS [53, 54].

Earlier it has been reported that flavoproteins, generally, can activate oxygen, resulting in the production of $O_2^{\bullet-}$ and/or H_2O_2 [55-58]. Lipoamide dehydrogenase, the E3 subunit of α–KGDH, is a flavoenzyme being responsible for the transfer of reducing equivalents from lipoate, which is bound to the E2 subunit, to NAD^+, the final electron acceptor in the α–KGDH catalysed reaction. The group of Tretter and Adam-Vizi [54] and Starkov et al., [53] have demonstrated that α–KGDH, during the physiological catalytic function of the enzyme, generates H_2O_2 and to a smaller extent $O_2^{\bullet-}$. H_2O_2 generation by α–KGDH is strongly inhibited in the presence of NAD^+ and accelerated by addition of NADH [54]. Relatively to these findings, it has been suggested that the most important regulator of the α–KGDH-mediated ROS generation is the high NADH/NAD^+ ratio, which is commonly related to nutrient overload. Remarkably, adipogenesis is commonly associated with a cellular positive energetic state (high NADH/NAD^+ ratio), thus suggesting a potential involvement of α–KGDH-mediated ROS generation in adipocytes maturation.

2.2. p_{66}^{SHC} Participates in Controlling Ros Production in Adipocytes

The generation of H_2O_2 by mitochondria during nutrient overload, however, is not just the by-product of metabolic flux but can also be the result of specific enzymatic activity, such as $p66^{Shc}$ [59]. $p66^{Shc}$ is a downstream target of the insulin signalling pathway and a critical mediator of insulin-dependent mitochondrial ROS up-regulation in adipocytes [60-61]. $p66^{Shc}$ functions as an inducible redox enzyme, which is activated by nutrient/energy excess and generates H_2O_2 [59]. For this function, $p66^{Shc}$ uses reducing equivalents of the ETC through the direct oxidation of cytochrome *c* [59]. The H_2O_2 generated by $p66^{Shc}$ is ~30% that of the total pool of intracellular H_2O_2 and is biologically relevant, as shown by $p66^{Shc}$ ability *in vitro* and *in vivo* to induce mitochondrial permeability transition pore and by the findings that fat cells and AT derived from $p66^{Shc}$ KO mice accumulate significantly less oxidative stress [62]. Moreover, $p66^{Shc}$ KO mice have reduced body weight due to reduced mass of both WAT and BAT. In addition, $p66^{Shc}$-KO mice are protected from diet-induced obesity, suggesting that $p66^{Shc}$ is a genetic determinant of fat development in adult mice. $p66^{Shc}$-generated H_2O_2 appears to regulate insulin signalling selectively and at multiple levels. In $p66^{Shc}$ KO adipocytes, activation of AKT by insulin is attenuated, whereas MAPK activation is normal. Selective activation of AKT by $p66^{Shc}$ might be due to a direct effect of $p66^{Shc}$-generated H_2O_2 on phosphatases that regulate PI3K, such as PTEN [63]. One implication of these findings is that $p66^{Shc}$ functions to connect mitochondrial respiration to insulin signalling. Although it has never been demonstrated, it is possible to suppose that also $p66^{Shc}$, together with other ROS producing systems, could be employed in endorsing pre-adipocytes transformation into mature adipocytes [60] (Figure 1C).

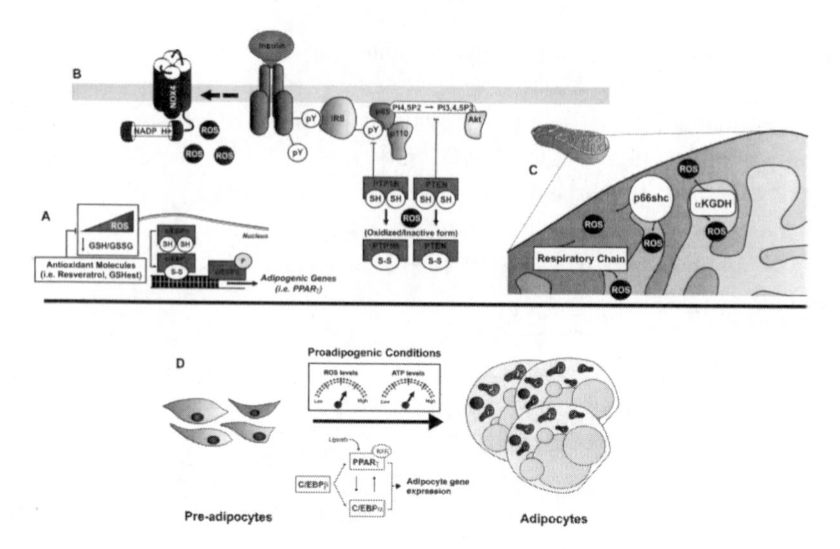

Figure 1. Schematic representation of the role of ROS in adipogenesis.

3. The Downside Of Ros in Adipogenesis

Obesity is a phenotypic alteration consequential to a chronic nutrient overload that, by producing a small amount of mitochondrial ROS, could accelerate pre-adipocytes differentiation. During the development of obesity, visceral AT bed expands and resident adipocytes become relatively hypoxic due to the inability of the vasculature to keep pace with AT remodelling. It has been reported that mitochondrial ROS production also rises under hypoxic conditions, and these ROS are involved in activation of hypoxia-inducible factors (HIFs), which mediate a transcriptional response that promotes adaptation to hypoxia [49, 64-66]. ROS production under hypoxic conditions occurs primarily at complex III, although the precise mechanisms involved are not clear yet. Interestingly, hypoxia or mitochondrial drugs seem to limit preadipocytes differentiation. Indeed, it has been demonstrated that mitochondrial toxins, by enhancing ROS, increase the anti-adipogenic CHOP-10/GADD153 levels, thus triggering hypoxia-dependent inhibition of adipocyte differentiation.

Beside hypoxia, hypo-vascularized AT could promote also a starvation of nutrients in resident adipocytes. However, although the mechanism by which nutrient overload could enhance mitochondrial ROS has been explained [34, 67], a low number of studies have eviscerated their origin in starving cells. It has been proposed that under these conditions, cells switch to a survival mode catabolizing "ready to use" cellular constituents for energy replacement. Proline is recycled as an energy source being promptly metabolized in Δ^1-pyrroline-5-carboxylate (P5C) by the mitochondrial inner membrane enzyme proline dehydrogenease or proline oxidase (PRODH/POX). The catalytic mechanism involves the transfer of electrons from substrate proline to FAD, with cytochrome *c* as the subsequent carrier into the ETC. Thus, proline is a direct substrate for the generation of ATP [68]. Interestingly, during nutrient lack, POX activation also modulates cellular redox homeostasis enhancing mitochondrial ROS, which act as mediators of several stress responsive pathways. Accordingly, nutrient restriction elicits POX induction in *C. elegans*, with consequent mitochondrial ROS production accompanied by metabolic adaptive responses [69-70]. However, nowadays there are no evidences about the role of POX in adipogenesis and in adipose cells.

Conclusion

Adipocytes are dynamic reservoirs for energy homeostasis in mammals. The cellular development associated with AT growth involves both cellular hypertrophy (increase in size) and hyperplasia (increase in number). Both adipocyte hypertrophy and hyperplasia occur in association with high nutrients levels. Hypertrophy is exclusively related to triglycerides accumulation into lipid droplets, whereas hyperplasia is an intrinsic process strongly related to pre-adipocytes differentiation into mature lipid-filled adipocytes. Although much progress has been made in understanding the transcriptional program that induces adipogenesis, the signalling mechanisms underlying the activation of transcriptional machinery are not fully understood.

A plethora of works confirmed the key role of suitable doses of ROS in promoting adipogenesis. In line with the above-described evidence, it can be suggested that a tight tuning of ROS producing systems participate in adipogenesis concomitantly to high nutrient levels (Figure 1D). In this scenario mitochondria likely represent the main ROS power plan, discerning a pro-adipogenic versus an anti-adipogenic function.

References

[1] Sun K, Kusminski CM & Scherer PE (2011) Adipose tissue remodeling and obesity. *J. Clin. Invest.* 12:2094-2101.

[2] Arner E, Westermark PO, Spalding KL, Britton T, Rydén M, Frisén J, Bernard S & Arner P (2010) Adipocyte turnover: relevance to human adipose tissue morphology. *Diabetes* 59:105-109.

[3] Spalding KL, Arner E, Westermark PO, Bernard S, Buchholz BA, Bergmann O, Blomqvist L, Hoffstedt J, Näslund E, Britton T, Concha H, Hassan M, Rydén M, Frisén J & Arner P (2008) Dynamics of fat cell turnover in humans. *Nature* 453:783-787.

[4] Rodeheffer MS, Birsoy K, & Friedman JM (2008) Identification of white adipocyte progenitor cells in vivo. *Cell* 135:240-249.

[5] Higuchi M, Dusting GJ, Peshavariya H, Jiang F, Hsiao ST, Chan EC & Liu GS (2013) Differentiation of human adipose-derived stem cells into fat involves reactive oxygen species and Forkhead box O1 mediated upregulation of antioxidant enzymes. *Stem Cells Dev.* 22:878-888.

[6] Lee H, Lee YJ, Choi H, Ko EH & Kim JW (2009) Reactive oxygen species facilitate adipocyte differentiation by accelerating mitotic clonal expansion. *J. Biol. Chem.* 284:10601-10609.

[7] Kanda Y, Hinata T, Kang SW & Watanabe Y (2011) Reactive oxygen species mediate adipocyte differentiation in mesenchymal stem cells. *Life Sci.* 89:250-258.

[8] Aquilano K, Baldelli S & Ciriolo MR (2011) Glutathione is a crucial guardian of protein integrity in the brain upon nitric oxide imbalance. *Commun.Integr.Biol.* 4:477-479.

[9] Vigilanza P, Aquilano K, Baldelli S, Rotilio G & Ciriolo MR (2011) Modulation of intracellular glutathione affects adipogenesis in 3T3-L1 cells. *J. Cell. Physiol.* 226:2016-2024.

[10] Kim JW, Tang QQ, Li X & Lane MD (2007) Effect of phosphorylation and S-S bond-induced dimerization on DNA binding and transcriptional activation by C/EBPbeta. *Proc. Natl. Acad. Sci. USA* 104:1800-1804.

[11] Chutkow WA, Birkenfeld AL, Brown JD, Lee HY, Frederick DW, Yoshioka J, Patwari P, Kursawe R, Cushman SW, Plutzky J, Shulman GI, Samuel VT & Lee RT. (2010) Deletion of the alpha-arrestin protein Txnip in mice promotes adiposity and adipogenesis while preserving insulin sensitivity. *Diabetes* 59:1424-1434.

[12] Lettieri-Barbato D, Tomei F, Sancini A, Morabito G & Serafini M (2013) Effect of plant foods and beverages on plasma non-enzymatic antioxidant capacity in human subjects: a meta-analysis. *Br.J.Nutr.* 109:1544-1556.

[13] Mitterberger MC & Zwerschke W (2013) Mechanisms of Resveratrol-Induced Inhibition of Clonal Expansion and Terminal Adipogenic Differentiation in 3T3-L1 Preadipocytes. Mar 22. *J. Gerontol A Biol. Sci. Med. Sci.*

[14] Kwon JY, Seo SG, Yue S, Cheng JX, Lee KW & Kim KH (2012) An inhibitory effect of resveratrol in the mitotic clonal expansion and insulin signaling pathway in the early phase of adipogenesis. *Nutr. Res.* 32:607-616.

[15] Vingtdeux V, Chandakkar P, Zhao H, Davies P & Marambaud P (2011) Small-molecule activators of AMP-activated protein kinase (AMPK), RSVA314 and RSVA405, inhibit adipogenesis. *Mol.Med.* 17:1022-1030.

[16] Kim S, Jin Y, Choi Y & Park T (2011) Resveratrol exerts anti-obesity effects via mechanisms involving down-regulation of adipogenic and inflammatory processes in mice. *Biochem.Bharmacol.* 81:1343-1351.

[17] Brunet A, Sweeney LB, Sturgill JF, Chua KF, Greer PL, Lin Y, Tran H, Ross SE, Mostoslavsky R, Cohen HY, Hu LS, Cheng HL, Jedrychowski MP, Gygi SP, Sinclair DA, Alt FW & Greenberg ME (2004) Stress-dependent regulation of FOXO transcription factors by the SIRT1 deacetylase. *Science* 303:2011-2015.

[18] Boily G, Seifert EL, Bevilacqua L, He XH, Sabourin G, Estey C, Moffat C, Crawford S, Saliba S, Jardine K, Xuan J, Evans M, Harper ME & McBurney MW (2008) SirT1 regulates energy metabolism and response to caloric restriction in mice. *PLoS One* 3:e1759.

[19] Picard F, Kurtev M, Chung N, Topark-Ngarm A, Senawong T, Machado De Oliveira R, Leid M, McBurney MW & Guarente L (2004) Sirt1 promotes fat mobilization in white adipocytes by repressing PPAR-gamma. *Nature* 429:771-776.

[20] Backesjo CM, Li Y, Lindgren U & Haldosen LA (2009) Activation of Sirt1 decreases adipocyte formation during osteoblast differentiation of mesenchymal stem cells. *Cells Tissues Organs* 189:93-97.

[21] Gwinn DM, Shackelford DB, Egan DF, Mihaylova MM, Mery A, Vasquez DS, Turk BE & Shaw RJ (2008) AMPK phosphorylation of raptor mediates a metabolic checkpoint. *Mol. Cell* 30:214-226.

[22] Chen S, Li Z, Li W, Shan Z & Zhu W (2011) Resveratrol inhibits cell differentiation in 3T3-L1 adipocytes via activation of AMPK. *Can. J. Physiol. Pharmacol.* 89:793-799.

[23] Giri S, Rattan R, Haq E, Khan M, Yasmin R, Won JS, Key L, Singh AK & Singh I. (2006) AICAR inhibits adipocyte differentiation in 3T3L1 and restores metabolic alterations in diet-induced obesity mice model. *Nutr. Metab. (Lond)* 3:31.

[24] Lee H, Kang R, Bae S & Yoon Y (2011) AICAR, an activator of AMPK, inhibits adipogenesis via the WNT/beta-catenin pathway in 3T3-L1 adipocytes. *Int. J. Mol. Med.* 28:65-71.

[25] Klemm DJ, Leitner JW, Watson P, Nesterova A, Reusch JE, Goalstone ML & Draznin B. (2001) Insulin-induced adipocyte differentiation. Activation of CREB rescues adipogenesis from the arrest caused by inhibition of prenylation. *J. Biol. Chem.* 276:28430-28435.

[26] Bluher M, Michael MD, Peroni OD, Ueki K, Carter N, Kahn BB & Kahn CR (2002) Adipose tissue selective insulin receptor knockout protects against obesity and obesity-related glucose intolerance. *Dev. Cell.* 3:25-38.

[27] Zhang HH, Huang J, Düvel K, Boback B, Wu S, Squillace RM, Wu CL & Manning BD (2009) Insulin stimulates adipogenesis through the Akt-TSC2-mTORC1 pathway. *PLoS One* 4:e6189.

[28] Goldstein BJ, Mahadev K, Wu X, Zhu L & Motoshima H (2005) Role of insulin-induced reactive oxygen species in the insulin signaling pathway. *Antioxid.Redox Signal.* 7:1021-1031.

[29] Mahadev K, Motoshima H, Wu X, Ruddy JM, Arnold RS, Cheng G, Lambeth JD & Goldstein BJ (2004) The NAD(P)H oxidase homolog Nox4 modulates insulin-stimulated generation of H2O2 and plays an integral role in insulin signal transduction. *Mol Cell Biol* 24:1844-1854.

[30] Mouche S, Mkaddem SB, Wang W, Katic M, Tseng YH, Carnesecchi S, Steger K, Foti M, Meier CA, Muzzin P, Kahn CR, Ogier-Denis E & Szanto I. (2007) Reduced expression of the NADPH oxidase NOX4 is a hallmark of adipocyte differentiation. *Biochim. Biophys. Acta.* 1773:1015-1027.

[31] Schroder K, Wandzioch K, Helmcke I & Brandes RP (2009) Nox4 acts as a switch between differentiation and proliferation in preadipocytes. *Arterioscler.Thrombo.Vasc.Biol.* 29:239-245.

[32] Martyn KD, Frederick LM, von Loehneysen K, Dinauer MC & Knaus UG (2006) Functional analysis of Nox4 reveals unique characteristics compared to other NADPH oxidases. *Cell Signal.* 18:69-82.

[33] Takac I, Schröder K, Zhang L, Lardy B, Anilkumar N, Lambeth JD, Shah AM, Morel F & Brandes RP (2011) The E-loop is involved in hydrogen peroxide formation by the NADPH oxidase Nox4. *J. Biol. Chem.* 286:13304-13313.

[34] Choi K & Kim YB (2010) Molecular mechanism of insulin resistance in obesity and type 2 diabetes. *Korean J. Intern. Med.* 25:119-129.

[35] Rhee SG (2006) Cell signaling. H2O2, a necessary evil for cell signaling. *Science* 312:1882-1883.

[36] Tonks NK (2006) Protein tyrosine phosphatases: from genes, to function, to disease. *Nat. Rev. Mol. Cell. Biol.* 7:833-846.

[37] Loh K, Deng H, Fukushima A, Cai X, Boivin B, Galic S, Bruce C, Shields BJ, Skiba B, Ooms LM, Stepto N, Wu B, Mitchell CA, Tonks NK, Watt MJ, Febbraio MA, Crack PJ, Andrikopoulos S & Tiganis T (2009) Reactive oxygen species enhance insulin sensitivity. *Cell Metab.* 10:260-272.

[38] Wijesekara N, Konrad D, Eweida M, Jefferies C, Liadis N, Giacca A, Crackower M, Suzuki A, Mak TW, Kahn CR, Klip A & Woo M (2005) Muscle-specific Pten deletion protects against insulin resistance and diabetes. *Mol. Cell. Biol.* 25:1135-1145.

[39] Elchebly M, Payette P, Michaliszyn E, Cromlish W, Collins S, Loy AL, Normandin D, Cheng A, Himms-Hagen J, Chan CC, Ramachandran C, Gresser MJ, Tremblay ML & Kennedy BP (1999) Increased insulin sensitivity and obesity resistance in mice lacking the protein tyrosine phosphatase-1B gene. *Science* 283:1544-1548.

[40] Kwon J, Lee SR, Yang KS, Ahn Y, Kim YJ, Stadtman ER & Rhee SG (2004) Reversible oxidation and inactivation of the tumor suppressor PTEN in cells stimulated with peptide growth factors. *Proc. Natl. Acad. Sci. USA* 101:16419-16424.

[41] Lee SR, Yang KS, Kwon J, Lee C, Jeong W & Rhee SG (2002) Reversible inactivation of the tumor suppressor PTEN by H2O2. *J. Biol. Chem.* 277:20336-20342.

[42] Meng TC, Buckley DA, Galic S, Tiganis T & Tonks NK (2004) Regulation of insulin signaling through reversible oxidation of the

protein-tyrosine phosphatases TC45 and PTP1B. *J. Biol. Chem.* 279:37716-37725.

[43] Mahadev K, Zilbering A, Zhu L & Goldstein BJ (2001) Insulin-stimulated hydrogen peroxide reversibly inhibits protein-tyrosine phosphatase 1b in vivo and enhances the early insulin action cascade. *J. Biol. Chem.* 276:21938-21942.

[44] Mahadev K, Wu X, Zilbering A, Zhu L, Lawrence JT & Goldstein BJ (2001) Hydrogen peroxide generated during cellular insulin stimulation is integral to activation of the distal insulin signaling cascade in 3T3-L1 adipocytes. *J. Biol. Chem.* 276:48662-48669.

[45] Barbato DL, Tomei G, Tomei F & Sancini A (2010) Traffic air pollution and oxidatively generated DNA damage: can urinary 8-oxo-7,8-dihydro-2-deoxiguanosine be considered a good biomarker? A meta-analysis. *Biomarkers* 15:538-545.

[46] Kusminski CM & Scherer PE (2012) Mitochondrial dysfunction in white adipose tissue. *Trends in endocrinology and metabolism: TEM* 23(9):435-443.

[47] De Pauw A, Tejerina S, Raes M, Keijer J & Arnould T (2009) Mitochondrial (dys)function in adipocyte (de)differentiation and systemic metabolic alterations. *Am. J. Pathol.* 175:927-939.

[48] Tormos KV, Anso E, Hamanaka RB, Eisenbart J, Joseph J, Kalyanaraman B & Chandel NS (2011) Mitochondrial complex III ROS regulate adipocyte differentiation. *Cell Metab.* 14:537-544.

[49] Guzy RD, Hoyos B, Robin E, Chen H, Liu L, Mansfield KD, Simon MC, Hammerling U & Schumacker PT (2005) Mitochondrial complex III is required for hypoxia-induced ROS production and cellular oxygen sensing. *Cell Metab.* 1:401-408.

[50] Droge W (2002) Free radicals in the physiological control of cell function. *Physiol. Rev.* 82:47-95.

[51] Wellen KE & Thompson CB (2010) Cellular metabolic stress: considering how cells respond to nutrient excess. *Molecular cell.* 40:323-332.

[52] Lettieri Barbato D, Baldelli S, Pagliei B, Aquilano K & Ciriolo MR (2012) Caloric Restriction and the Nutrient-Sensing PGC-1alpha in Mitochondrial Homeostasis: New Perspectives in Neurodegeneration. *Int. J. Cell. Biol.* 2012:759583.

[53] Starkov AA, Fiskum G, Chinopoulos C, Lorenzo BJ, Browne SE, Patel MS & Beal MF (2004) Mitochondrial alpha-ketoglutarate

dehydrogenase complex generates reactive oxygen species. *J. Neurosci.* 24:7779-7788.

[54] Tretter L & Adam-Vizi V (2004) Generation of reactive oxygen species in the reaction catalyzed by alpha-ketoglutarate dehydrogenase. *J. Neurosci.* 24:7771-7778.

[55] Ballou D, Palmer G & Massey V (1969) Direct demonstration of superoxide anion production during the oxidation of reduced flavin and of its catalytic decomposition by erythrocuprein. *Biochem. Biophys. Res. Commun.* 36:898-904.

[56] Chan PC & Bielski BH (1980) Glyceraldehyde-3-phosphate dehydrogenase-catalyzed chain oxidation of reduced nicotinamide adenine dinucleotide by perhydroxyl radicals. *J. Biol. Chem.* 255:874-876.

[57] Kakinuma K, Fukuhara Y & Kaneda M (1987) The respiratory burst oxidase of neutrophils. Separation of an FAD enzyme and its characterization. *J. Biol. Chem.* 262:12316-12322.

[58] Massey V (1994) Activation of molecular oxygen by flavins and flavoproteins. *J. Biol. Chem.* 269:22459-22462.

[59] Giorgio M, Migliaccio E, Orsini F, Paolucci D, Moroni M, Contursi C, Pelliccia G, Luzi L, Minucci S, Marcaccio M, Pinton P, Rizzuto R, Bernardi P, Paolucci F & Pelicci PG (2005) Electron transfer between cytochrome c and p66Shc generates reactive oxygen species that trigger mitochondrial apoptosis. *Cell* 122:221-233.

[60] Berniakovich I, Trinei M, Stendardo M, Migliaccio E, Minucci S, Bernardi P, Pelicci PG & Giorgio M (2008) p66Shc-generated oxidative signal promotes fat accumulation. *J. Biol. Chem.* 283:34283-34293.

[61] Tomilov AA, Bicocca V, Schoenfeld RA, Giorgio M, Migliaccio E, Ramsey JJ, Hagopian K, Pelicci PG & Cortopassi GA (2010) Decreased superoxide production in macrophages of long-lived p66Shc knock-out mice. *J. Biol. Chem.* 285:1153-1165.

[62] Francia P, delli Gatti C, Bachschmid M, Martin-Padura I, Savoia C, Migliaccio E, Pelicci PG, Schiavoni M, Lüscher TF, Volpe M & Cosentino F (2004) Deletion of p66shc gene protects against age-related endothelial dysfunction. *Circulation* 110:2889-2895.

[63] Tang X, Powelka AM, Soriano NA, Czech MP & Guilherme A (2005) PTEN, but not SHIP2, suppresses insulin signaling through the phosphatidylinositol 3-kinase/Akt pathway in 3T3-L1 adipocytes. *J. Biol. Chem.* 280:22523-22529.

[64] Brunelle JK, Bell EL, Quesada NM, Vercauteren K, Tiranti V, Zeviani M, Scarpulla RC & Chandel NS (2005) Oxygen sensing requires mitochondrial ROS but not oxidative phosphorylation. *Cell Metab.* 1:409-414.

[65] Guzy RD & Schumacker PT (2006) Oxygen sensing by mitochondria at complex III: the paradox of increased reactive oxygen species during hypoxia. *Exp. Physiol.* 91:807-819.

[66] Mansfield KD, Guzy RD, Pan Y, Young RM, Cash TP, Schumacker PT & Simon MC (2005) Mitochondrial dysfunction resulting from loss of cytochrome c impairs cellular oxygen sensing and hypoxic HIF-alpha activation. *Cell Metab.* 1:393-399.

[67] Dirkx E, Schwenk RW, Glatz JF, Luiken JJ & van Eys GJ (2011) High fat diet induced diabetic cardiomyopathy. *Prostaglandins Leukot. Essent. Fatty Acids* 85:219-225.

[68] Hagedorn CH & Phang JM (1986) Catalytic transfer of hydride ions from NADPH to oxygen by the interconversions of proline and delta 1-pyrroline-5-carboxylate. *Arch.Biochem.Biophys.* 248:166-174.

[69] Zarse K, Schmeisser S, Groth M, Priebe S, Beuster G, Kuhlow D, Guthke R, Platzer M, Kahn CR & Ristow M (2012) Impaired insulin/IGF1 signaling extends life span by promoting mitochondrial L-proline catabolism to induce a transient ROS signal. *Cell Metab.* 15:451-465.

[70] Schroeder EA & Shadel GS (2012) Alternative mitochondrial fuel extends life span. *Cell Metab. 1*5:417-418.

In: Adipogenesis
Editors: Y. Lin and X. Cai

ISBN: 978-1-62808-750-5

Chapter VI

Adipogenesis: Signaling Pathways, Molecular Regulation and Clinical Impact

***Kristin A. McPhillips*[1] *and Quanjun Cui*[2,*]**

[1]Department of Surgery,
University of Virginia School of Medicine,
Charlottesville, Virginia, US
[2]Department of Orthopaedic Surgery,
University of Virginia School of Medicine,
Charlottesville, Virginia, US

Abstract

There has been an overwhelming amount of research in adipogenesis over the past several years and it has been cited as one of the most widely studied areas of cellular biology. The interest likely stems in part from the availability of in vitro models that seem to reliably replicate critical aspects of adipogenesis in vivo. The pathways from mesenchymal stem cell to preadipocyte to adipocyte are well defined. Key mediators in stem

* Corresponding author: Quanjun Cui, MD, MSc. Department of Orthopedic Surgery, University of Virginia School of Medicine, Charlottesville, VA 22908. Phone: 434-243-0236, Fax: 434-243-0242. E-mail: QC4Q@hscmail.mcc.virginia.edu.

cell commitment have been identified in the BMP and Wnt families. Fat genes, notably PPAR-γ, have been identified, and regulators of the expression of the adipocyte phenotype in committed preadipocytes have also been identified all along the pathway, from epigenetic histone gene modulators to modifiers of protein folding in the endoplasmic reticulum. There is an increasing focus on the effect of inflammatory markers like IGF-1 and TNF, as well as environmental toxins and exposures, such as organophosphates, arsenic and obesogen exposures in-utero. Compounds like ginseng, ginkgo, selenate and resveratrol are showing promise as down-regulators of adipogenesis. Obesity and its sequelae--diabetes, hypertension and vascular disease, are global epidemics, and further understanding of the regulation of this pathway has the potential to have a profound clinical impact. In addition to its relevance for obesity and the metabolic syndrome, dysregulation of adipogenesis has been implicated in many other clinical problems, from osteoporosis to aging and cancer. Although research on a cellular level and in animal models has shown promise, it has yet to carry over to the bedside. Continued research is needed to further elucidate the roles and modulators of each pathway in order to develop drugs that either reduce adipogenesis or make adipose tissue more thermogenic.

Introduction

When Odysseus and his men land on the island of the Sun God, Odysseus is warned that neither he nor his crew should eat the Sun God's cows, lest his ship and crew be destroyed. Eurylochus, his first mate, convinces the men to eat the cows, arguing "if he [the Sun God] be somewhat wroth for his cattle with straight horns, and is fain to wreck our ship, and the other gods follow his desire, rather with one gulp at the wave would I cast my life away, than be slowly straitened to death in a desert isle " (Homer 850 BC).

Fat has long been recognized as a storage vessel for excess caloric intake, and until relatively recently it was universally prized, due to unpredictable food supplies and the constant specter of starvation. When Abel gave God a sheep and the fat of the sheep, his gift is respected; when his brother offers God the fruits of the earth his gift is not respected (Genesis 4:2). There are also references to the "fat of the earth," and the righteous, who will "flourish and be fat" (Psalm 92:12-14). The souls of the diligent and the liberal are fat (proverbs 13:4, 11:25) and he who trusts in the Lord is also made fat (proverbs 28:25). When the prodigal son returns home his father kills not just any cow, but a "fatted calf" (Luke 15:27). [1] Until relatively recently, fat has been perceived as an indication of health and prosperity; it is speculated that fat may have conferred resistance to tuberculosis and other infectious diseases that were epidemic in the 1800s [2].

However, as early as the 1940's it was noted that obesity was associated with cardiovascular disease, [3] and Hippocrates noted in his aphorisms that obesity was associated with a higher risk of sudden death [4].

The earliest reports of adipose tissue in the literature are from the 1500s from Swiss naturalist Konrad Gessner [5]. Most early research sought to prove the existence of a specialized fat cell and characterize its nature. The predominant method was to starve animals and then examine the remaining tissue, with the hope that the fat would be depleted so that the true nature of the cell would be revealed; unfortunately, all of the animals died before their fat reserves were completely depleted [6]. Fat tissue was long recognized as a storage vessel for excess caloric intake, but the nature of the cells that make up the tissue and the way that the tissue functions, remained and to some extent remains a mystery. Koliker, in the 1800s, described primitive glands produced from special kinds of connective tissue [7]. There are also reports that fat was a part of connective tissue and could be laid down anywhere in cases of excessive feeding [8, 9] and of fat-forming organs [10]. Waldeyer described a wandering fat cell, whereas most other reports describe a fixed cell [11]. Shaw described a likeness to epithelial cells because of the abundance of protoplasm [6]. The concept of adipose tissue being related to or being a part of the reticulo-endothelial system was proposed by Wasserman; this was in harmony with early observations that the fat cells appeared to be glands and had affinity for blood cells. He thought they could re-establish their blood forming function if depleted [6].

Other than these initial studies attempting to discover the nature of fat cells, and a few studies about the location of fat deposits, adipose tissue was largely ignored in research and adipose cells were viewed as a mostly inert storage vessel for lipids [12] The notion of adipose tissue as an active participant in energy homeostasis has emerged over the past two decades. In 1987, it was discovered that adipose tissue was a major site of metabolism of sex steroids [13], and in 1994 leptin, a cytokine-like factor secreted from adipose tissue, was discovered [14]. Many novel proteins secreted from adipocytes known as adipokines have been identified, and they have been shown to play a role in energy homeostasis, inflammation and immune function [15, 16].

No longer just a storage vessel for triglycerides, adipose tissue is now widely viewed as an endocrine organ [17]. Adipocytes develop during gastrulation from the mesoderm layer, the same layer from which myocytes and chondrocytes develop [18]. Two histologically distinct types of cells are formed; white fat, found mostly in areas rich in connective tissue between the muscle and dermis and around internal organs, and brown fat, found between the scapulae and surrounding the great vessels. [19-21]. The mechanism of energy dissipation by brown fat is the uncoupling of respiration from ATP production via uncoupling protein 1 (UCP1) [22]. Structurally, white adipose cells possess lipids organized in one large unilocular droplet with the remainder of cell contents compressed to the periphery, whereas brown adipocytes are composed of fat stored in multilocular droplets, and are also characterized by a high content of mitochondria and a central nucleus. Physiologically, white adipose tissue is specialized in energy storage, and brown adipose tissue regulates thermogenesis in response to food intake and cold.

In Vitro Models of Adipogenesis

Since the 1970s, there has been an exponential increase in the volume of research in the field of adipogenesis, [23]. Taylor and Jones showed that the mesodermal progenitor cells that give rise to adipose tissue can also give rise to muscle, cartilage, and fat [24]. The first preadipose cell lines were developed in the 1970s by Green and colleagues [25]. Although a number of other cell lines now exist, the 3T3-LI and 3T3-F442A murine preadipocyte cell lines are still the most widely studied cell lines [26]. These cells are morphologically indistinguishable from fibroblasts, although they are committed to the adipocyte lineage. They have undergone differentiation, but are arrested in various stages of development. The cell lines are monitored for terminal differentiation with use of oil red staining, and are monitored for proteins expressed by adipocytes. Since then, several other preadipocyte cell lines have also been studied; they differ in their optimal differentiation regimen, but overall behave very similarly to the original lines [26, 27].

Early studies began with stem cells isolated from the bone marrow stromal cells of mice [26]; Lanotte et al. showed that cells from both epithelioid and fibroid morphologies had a high frequency of differentiating into adipocytes, and that insulin insensitivity, as well as inhibition of mitosis, were markers of terminal differentiation [28]. Cells obtained were ultrastructurally similar to those grown in long term culture, indicating that in-vitro models would be good models for in-vivo processes [26]. The majority of research on adipogenesis is based on in-vitro models from preadipocytes, namely cells that are committed to the adipocyte lineage but not yet fully differentiated. The events that promote the determination of mesenchymal stem cells to the adipose cell lineage and the markers of cell determination remain poorly understood. There are several barriers to the study of the pluripotent stem cells within adipose tissue. Because adipose tissue is very diffuse in vivo, and because undifferentiated precursor cells are difficult to culture and are relatively small in number in comparison to the rest of the cells in the tissue, studies of embryonic stem cells are a poor model for in-vivo influence.

There is a growing body of literature about the early determinants of mesenchymal stem cell fate. Recently, there has been great attention paid to the bone morphogenic protein family and its role in regulating adipogenesis, as well as the relationship between osteogenesis, hematopoiesis and adipogenesis [29]. Paired homeobox transcription factors, Prrx1a, Prrx1b, and Prrx2, have also been implicated in cell fate as negative regulators of adipogenesis [30].

Once the cell is committed to the pre-adipocyte lineage, the first step in adipogenesis is growth arrest, which is normally achieved through contact inhibition. However, contact inhibition is not an absolute prerequisite for adipogenesis, as cells plated at low density in a serum-free medium [31] or methylcellulose suspension [32] have been shown to undergo differentiation.

After growth arrest the cells undergo induction--the steps that lead to the acquisition of adipocyte characteristics. The conditions for the initiation of induction vary according to cell cultures, but generally induction commences in the presence of insulin, dexamethasone and isobutylmethylxanthine. Along with morphogenic changes, these cells were noted to have widespread alteration in protein composition as well as changes in the patterns of secreted proteins [33]. Relatively early on in the understanding of the metabolic role of adipose, insulin and glucocorticoid were noted to accelerate the phenotypic changes in the adipocyte [34]. A study by Harrison et al. showed that induction can occur in a serum-free medium containing insulin, indicating that insulin is the only necessary inducing factor [35]. In addition, several other studies reiterated the role of glucocorticoids in induction [36, 37].

The first hallmark of adipogenesis is a conversion in cell shape from fibroblastic to spherical. The morphologic changes are caused by changes at the level of the extracellular matrix and cytoskeletal components. [31]. The processes that regulate cytoskeleton remodeling are still poorly understood, but are beginning to receive more attention; one recent study proposed that the first step is the upregulation of an alphatubulin acetylase, which initiates cytoskeleton remodeling [38]. Several studies have shown that extracellular stresses [39] and extracellular adhesion molecules [40] influence mesenchymal cell fate. SPARC (secreted protein acidic and rich in cysteine) is a matrix protein that has generated a multitude of research questions over the past decade. It belongs to a family of matricellular proteins that regulate interactions between extracellular matrix and cells. It is antiadhesive in vitro and regulates collagen production in vivo. It has been shown to participate in tumor progression, healing and inflammation [41]. SPARC-null mice were shown to accumulate significantly more fat than wild-type mice [42]. Likewise, SPARC-null bone marrow showed a greater tendency to differentiate into adipocytes [43].

Morphologic changes are followed by one or two more rounds of replication known as clonal expansion, characterized by the induction of peroxisome proliferator-activated receptor y (PPARγ) and CCAAT/enhancer binding protein alpha (C/EBPα) [44, 45]. Following induction, there is a second period of growth arrest, which is followed by expression of the fully differentiated phenotype.

The mechanism by which PPARγ and C/EBPα bring about change remains uncertain. However, the shift from preadipocyte to fully differentiated cell is associated with changes in expression of many proteins that provide clues to the mechanism. Activation of PPARγ in non-confluent cells has been associated with a loss of DNA binding activity of E2F/DP, a transcription factor involved in the regulation of cell growth [46]. Several cyclin-dependent kinase inhibitors including p18, p21 and p27 were identified by Morrison and Farmer [45]. C/EBPα has also been shown to have antimitotic properties. Inhibition of mitosis blocks differentiation in cultured cell lines; however, this is not absolute, primary human adipocytes were shown to differentiate normally after being pretreated with alkylating agents that are powerful inhibitors of mitosis [47].

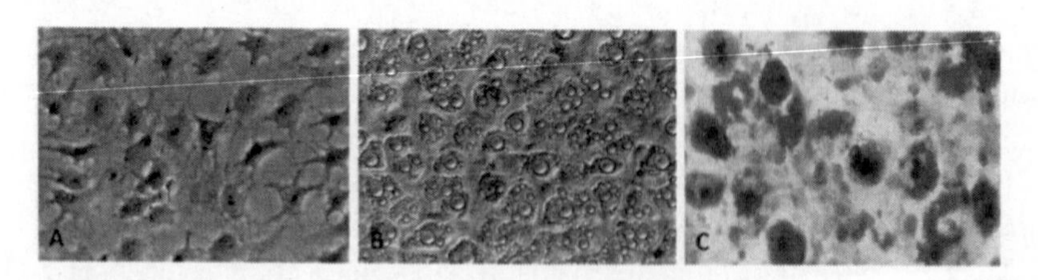

Figure 1A-C. Phase contrast micrographs of bone marrow derived mesenchymal stromal cells (A) treated with regular culture medium, (B) with addition of dexamethasone showing accumulation of lipid vesicles, and (C) stained with Sudan IV showed positive staining of the triglyceride vesicles.

After clonal expansion there is a second and final period of growth arrest. Studies that have tried to classify proteins by the timing of their expression, designating mRNAs into an early, intermediate and late expression groups, have fallen out of favor. Rosen and Spiegleman point out that in vitro studies are not ideal in the investigation of this expression timing because fat differentiation in culture occurs in clumps, and because the accumulation of an mRNA is dependent on both its rate of formation and degradation, which are unknown in most cases [26]. Still, it is known that the first events of adipogenesis are followed by C/EBPβ and C/EBPλ [48] followed by PPARγ and CEBPα, and then de novo expression of the adipocyte phenotype--including accumulation of triglycerides, expression of insulin receptors (glut 4), of aP2, fatty acid synthase and acetyl CoA-carboxylase [49].

Monitoring for terminal differentiation is accomplished by oil red O staining (Figure 1) and also by monitoring for the expression of adipocyte markers-- pre-adipocyte factor 1, lipoprotein lipase, adipocyte lipid binding protein, peroxisome proliferator receptor gamma (PPARγ), among others [27]. In-vitro models have many similarities to in-vivo models, including triglyceride storage, insulin sensitivity, and expression of adipocyte-specific genes [26]. However, these models are limited in several significant ways. Differentiation is incomplete and results in multilocular cells, whereas white adipocytes in-vivo consist of a single lipid drop; additionally, in-vitro adipocytes express less TNFα and leptin than do in-vivo adipocytes [50]. Interestingly, cultured cells do produce leptin if they are transplanted under the skin of an athymic mouse [50]. This indicates that either cultured cells do not accumulate as much fat as in vivo adipocytes, or there exists an extra-adipose factor that is not found in culture. These models do not allow for the study of adipogenesis in the context of the extracellular matrix and supporting structures, which have recently shown significance [51, 52]. Furthermore, adipose tissue arises in various locations--subcutaneous, retroperitoneal, and perivisceral-- and preadipocytes isolated from different sites have different potential, the basis for which is not well understood [53-55]. The behavior of mature cells also varies, which is important because there are known differences between visceral and subcutaneous fat pads [56-58]. These differences have clinical relevance: increased visceral adiposity is associated with a higher risk of metabolic syndrome (insulin resistance, hyperlipidemia, and cardiovascular disease) than is subcutaneous adiposity [59].

In Vivo Models of Adipogenesis

As in other areas of biology, in-vivo models of adipogenesis are based on transgenic or knockout models, which allow us to study adipogenesis under a variety of conditions. Nutritional models in which mice are fed high calorie diets provide models for studying the stages of adipose tissue development. Through these models, mice have been shown to have variable tendencies toward obesity and diabetes. Studies of effect modifiers have lead to study of the genes that confer either a predisposition to obesity or resistance, as well as environmental factors--like gut microbiota and specific diet components that affect adipogenesis [27].

Monogenic mouse obesity models include the leptin (ob/ob) or leptin receptor (db/ db) deficient mice. In humans, eight genes have been identified that are responsible for monogenic cases of obesity: leptin, leptin receptor, proconvertase 1, melanocortin receptor-4 and pro-opiomelan- ocortin (which are all involved in leptin signaling mediated appetite regulation) and Sim1, brain-derived neurotrophic factor and NTRK2 (which contribute to neurodevelopment) [27]. Mutations in these genes account for only 5% of patients with severe obesity [60]. The increasing prevalence of obesity can be in part attributed to a more sedentary lifestyle and wide access to a high calorie diet, but individual susceptibility to obesity is varied. Polygenic mouse models were developed to try to understand the genetic contribution to human obesity; these include the New Zealand obese mouse, the Tsumura Suzuki obese diabetes mouse, the M16 mouse, the Kuo Kondo mouse and the spiny mouse [27]. In humans, a total of 43 loci have been suggested to predispose to overall adiposity and 18 loci to visceral fat accumulation. Among them, fat mass and obesity associated (FTO) gene remains the best-replicated obesity gene. Still, there is much to be learned as far as the genetic determinants of adiposity; variations in the loci that have been identified so far so not correlate well with variations in BMI. [61].

Transcriptional Regulation and Signaling Pathways

The differentiation of white adipose cells from pre-adipocyte precursors has been studied extensively using determined cell lines (Figure 2 [62]).

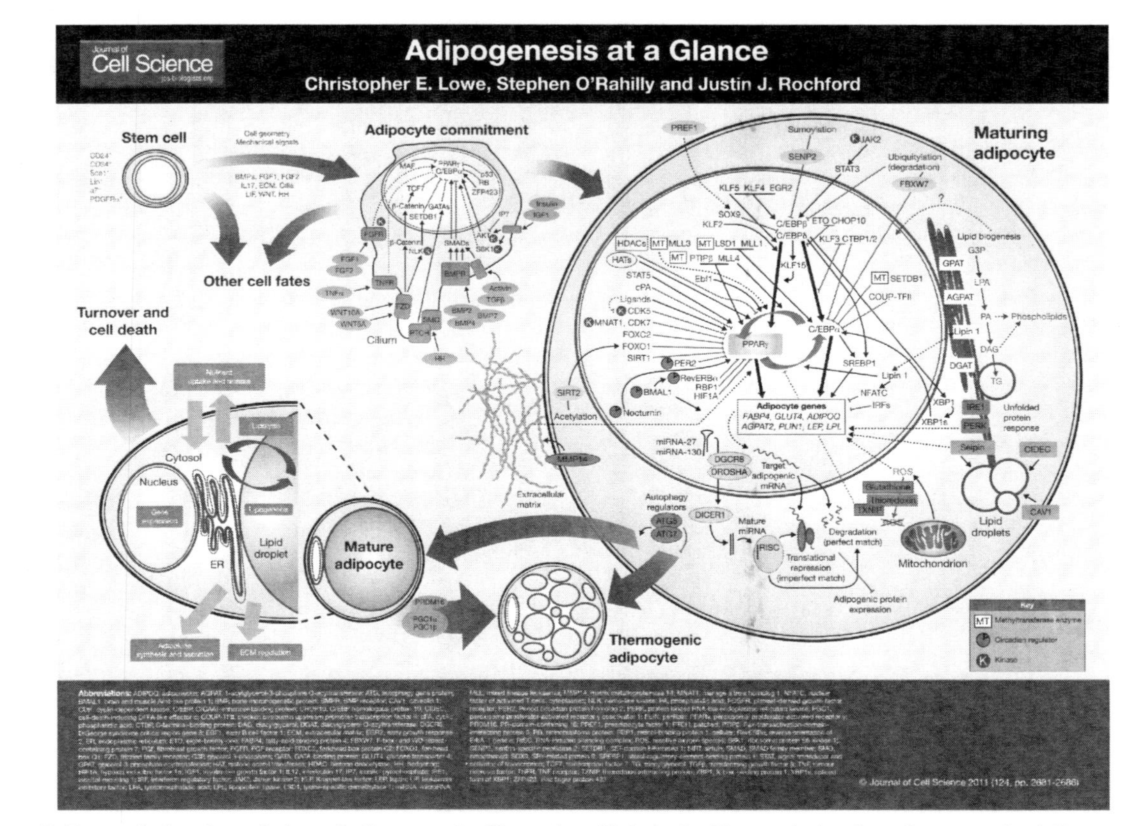

Figure 2. Transcriptional regulation of adipogenesis. (Reproduced/adapted with permission from Lowe, et al., Adipogenesis at a glance. Journal of Cell Science, 124, 2681-2686 [62]).

The most studied transcription factor is PPARγ; it belongs to a superfamily of ligand-activated transcriptors, which are important for controlling cellular processes, including lipid and carbohydrate metabolism [63]. PPARγ was first recognized as a necessary and sufficient modulator of adipogenesis in 1992 [64]. It is a heterodimer consisting of PPARγ and RXR [65]. It exists in 2 isoforms, which are created by alternative promoter usage and alternative splicing; PPARγ2 is highly fat-selective compared to PPARγ1, which is expressed in many tissues [65, 66] and it contains 30 additional amino acids. The structure of the PPARγ-RXR-DNA complex has recently been illustrated [67]; like many non-steroid nuclear receptors, it functions as an obligate heterodimer with RXR [68]. There has been no factor thus far that has been found to rescue adipogenesis in the absence of PPARγ. The homozygous null mutation is lethal secondary to placental insufficiency [69]. There is a report of one homozygous mutant that survived to term; the mouse lacked significant brown fat died shortly after birth [69]. Rosen et al. showed that cells null for PPARγ were excluded from white adipose tissue but not other tissues [70].

PPARγ has also been shown to be important in the *in vivo* models. Mice that did not express PPARγ2 and had altered PPARγ1 developed no white fat and very little brown fat, and had severe lipodystrophy, metabolic abnormalities, and early death. [69]. Studies have also shown that ablation of PPARγ in mature adipocytes leads to death and repopulation by PPARγ positive preadipocytes. [71]. PPARγ mutations have also been linked to human disease; few mutant genes have been identified, but those that have been identified have confirmed that PPARγ is essential for adipogenesis in humans [72, 73]. A PRo115 mutation that causes ligand-independent activation was found in four very obese subjects [74]. A more common mutation resulting in a less active PPAR-y is associated with lower BMI [75-77]. Mutations have been associated with familial lipodystrophies [78]. PPARγ is the target of antidiabetic thiazolidinediones [79]. Thiazolidinediones were first reported as antidiabetics in the 80s [80-82]. The mechanism of action of the thiazolidinediones was thought to be at post-receptor process [82]. Thiazolidinedione is a PPARγ agonist and has been shown to have beneficial properties for cardiovascular disease, with patients showing improvements in endothelial function, lipid profiles, and atherosclerosis, in addition to its role in the treatment of diabetes, as an insulin sensitizer. Thiazolidinedione has also been a potential target for liposarcoma treatment [83]. It has been shown to repress macrophages [84, 85] and it is suggested that it shows anti-inflammatory effects in several disease models including atherosclerosis [86], obesity-induced beta-cell dysfunction [87], multiple sclerosis and autoimmune encephalomyelitis [88], psoriasis [89], spinal cord injuries [90], inflammatory bowel disease [91], and arthritis [92].

The endogenous PPARγ ligand remains unknown. A ligand displacement assay has been used to look for a natural ligand for PPARγ; polyunsaturated fatty acids such as oleate and linoleanate have been shown to bind PPARγ, [93, 94] as have some eicosanoids [95]. These ligands bind within the range of Kd=2-50 uM, below most nuclear hormone receptors and over 1000 times below the affinity of the thiazolidinediones (40-700nM) [96]. A genome-wide analysis of PPARγ binding in mouse adipocytes identified over 5000 binding sites, many of which are also bound by RXRα [97, 98]. PPARγ binds target genes in the absence of ligand and in its ligand-free state it binds corepressors and silencing mediators, which recruit histone deacetylases and other chromatin-modifying enzymes that repress transcription. In some genes, PPARγ is constitutively associated with activators leading to high levels of transcription, and in others an exogenous ligand is needed to displace the repressor. Some steroid receptor coactivators have been found to be essential [99]; repressors have also been identified [100].

The second most studied transcription factor is the CAAT enhancer binding protein (C/EBP) family, a member of the leucine zipper class, which is seen in adipogenesis [101, 102]. In cultured cells, C/EBP-delta levels rise early and transiently after induction. [48, 103]. C/EBP-delta knockout mice fail to accumulate white or brown fat, have liver failure and die prematurely [104] even when the liver is rescued [105]. Mice that were lacking C/EBP-beta or alpha showed slight reduction in adipogenesis; they had normal white adipose tissue, but their brown adipose tissue showed reduction in lipid accumulation and UCP-1 expression [106]. Eighty-five percent of the mice that lacked both C/EBP-beta and alpha died in the perinatal period and the remaining 15% had reduced lipid accumulation in brown fat and reduced cell number, but normal cells in the white fat [106]. Ectopic expression of C/EBP beta and delta precede expression of PPARγ, and ectopic expression of C/EBP leads to induction of PPARγ [107]. C/EBP alpha is induced early and has been shown to colocalize with PPARγ [97]. Although C/EBP alpha knockout mice have dramatically reduced fat pads and metabolic derangements that lead to death in the first week of life from hypoglycemia [104], ectopic PPARγ can rescue adipogenesis in these models [70]. The exact mechanism of PPARγ-C/EBP cooperation has yet to be elucidated.

The Wnt signaling family is composed of lipid-modified glycoproteins that act by binding cell surface receptors and regulating gene expression via beta-catenin (canonical) or non-canonical pathways; these signals are active in a variety of development pathways and were first linked to adipogenesis in 2000 in a study by Ross et al. showing that Wnt induction in preadipocytes impaired adipocyte differentiation, and that inhibition of this pathway can lead to adipocyte-like characteristics in myoblasts [108]. Deletion of beta-catenin from mouse uterine tissue lead to the replacement of uterine muscle with fat cells [109]. Wnt has been shown to stabilize beta-catenin and prevent differentiation of preadipocytes. In bone, it has been shown to shift stromal progenitors from adipogenesis towards adipogenesis [110]. Regulation of the Wnt pathway is complex. A profile of the microRNAs that modulate the Wnt pathway revealed 29 miRNAs that activate Wnt pathways, which thus have the potential to downregulate adipogenesis, and 18 miRNAs that depress Wnt, and may upregulate adipogenesis [111]. There is also evidence that Wnt is important for adipogenesis in-vivo; humans with a Wnt 5a mutation have been shown to have a predisposition to obesity [21].

There are many other transcription factors that seem to play a role in adipocyte differentiation. A family of zinc-finger transcription factors, the KLFs, has also been recently implicated in adipogenesis [112]. They can act as activators or repressors, depending on the context. KLF4 is induced during the first 24 hours of differentiation and it transactivates C/EBPβ in collaboration with Krox 20 [112]. GATA binding proteins, which are found in preadipocytes in high numbers, seem to repress differentiation via repression of PPARγ [113]. Repression of GATA-2 has been associated with aplastic anemia [114]. Several interferon regulatory proteins have been shown to repress differentiation by binding promoters. C10orf116 has been recently indicated as a regulator of early differentiation of adipose cells; it was recently shown to increase insulin-stimulated glucose uptake in adipocytes via an increase in glut 4 levels, and it was shown to decrease apoptosis in preadipocytes [115].

Circadian rhythms have also been shown to have a role in the process via clock proteins [116, 117]. Timing is increasingly recognized as an important regulator of metabolism. Humans have been shown to have alternating cycles of lipogenesis and lipolysis. During the night, there is a predominance of lipolytic activity, which reduces hunger signaling and the need for food, whereas during the day lipogenesis predominates. [118] It is speculated that there are also circadian genes responsible for adipogenesis. Circadian rhythms have been established for the plasma concentrations of adipokines [119]. Melatonin significantly increased the expression of both CCAAT/EBPα and PPARγ in 3T3-L1 cells [120].

Several recent studies highlight the importance of histone methylation [121] and it has been shown that different demethylases and methyl transferases are active during adipogenesis [122, 123] Studies of genome wide DNAse activation have shown that chromatin remodeling long precedes gene activation [123]. Siersback et al. showed early binding of transcription factors like glucocorticoid, C/ERP-delta, retinoid X receptor, and Signal Transducer and Activator of Transcription 5a [124]. The binding of these transcription factors has been found to be necessary for the transcription of PPARγ [125].

Free Radicals

Free radicals and oxidative stress have recently been shown to reduce adipogenesis through several pathways. Free radicals reduce the C/EBP DNA-binding activity [126]. Radicals and oxidative stress from mitochondrial processes have been linked to signal transduction inducers, such as mitogen-activated protein kinase (MAPK) and serine kinases like Akt, which promote phosphorylation of hypoxia inducible factor-alpha (HIF-α) [127, 128], which represses PPAR-γ [129]. Hypoxia has also been shown to increase free radical formation by inhibiting mitochondrial electron transport, which also inhibits adipose differentiation via phosphorylation of HIF-α [130]. Hypoxia also results in a decrease in insulin sensitivity [131, 132]. In obese adipose tissue, it has been shown that HIF-1 levels increase but VEGF does not [133], resulting in reduced capillary density [132, 133]. Reactive oxygen species (ROS) have also been shown to affect mature adipocytes, resulting in insulin resistance via the down-regulation of GLUT-4 expression [134]. Galinier et al. demonstrated that adipose tissue from an obese rat model showed a higher content of glutathione and vitamin C in a lower redox state than the fat of lean animals [135], indicating that obesity is associated with reduced free radical formation. On the other hand, there are studies showing that lower levels of plasma antioxidants are associated with obesity prevalence [136]. Oxidative stress has also been associated with visceral fat accumulation and the metabolic syndrome [137] and, recently the oxidative stress was shown to precede the onset of insulin resistance and diabetes [138]. Race has been noted to modify the effect of oxidative stress on insulin sensitivity [139], possibly due to proinflammatory cytokines [140]. On a cellular level, inflammatory cytokines have been shown to be antiadipogenic [33, 141-143]. A recent study showed that nuclear factor E2 related factor (Nrf2), a transcription factor important for antioxidant and detoxification genes, was diminished during adipocyte differentiation [144]. In one mouse model, whether Nrf2 was knocked out globally, or specifically in adipocytes, reduced white adipose tissue and a severe metabolic syndrome with insulin resistance, hyperglycemia and hypertriglyceridemia was produced [145]. In another study, Nrf2 knockout was associated with steatohepatitis and impaired adipogenesis in white adipose tissue [146]. It has been suggested that changes in the ubiquitin-proteasome pathway that occur during the differentiation of 3T3-L1 cells decrease adipose differentiation and increase oxidative stress on mature adipocytes, promoting the deleterious effects on adipose aging [147].

Several antioxidants, among them oleanic acid [148], blueberry polyphenols [149], phytoestrogens [150], an extract from ginger [151], and purple potato extract [152] have shown promise in suppressing adipogenesis in cell cultures. Buckwheat was shown to decrease adipogenesis [153]; fermented rice bran was shown to inhibit ROS generation and upregulate PPARγ, adiponectin and glut4, increasing insulin sensitivity and neutralizing free radicals [154]. Brazilein inhibited intracellular lipid accumulation during adipocyte differentiation in 3T3-L1 cells and suppressed the induction of PPARγ [155]. Hydroxypropyl methyl cellulose, a nonfermentable soluble dietary fiber was shown to reduce weight gain and enhance insulin sensitivity and down regulate genes related to inflammation and adipogenesis in a mouse model [156]. Phytoestrogens like resveratrol have shown promise as down regulators of adipogenesis by upregulating the expression of osteo-lineage genes RUNX2 and osteocalcin, while suppressing adipo-lineage genes PPARγ2 and LEPTIN [157-159]. Phytoestrogens have also been shown to induce apoptosis and lipid mobilization in adipocytes [160]. Green tea has been shown to inhibit proliferation and induce apoptosis of adipocytes [161], it has also been shown to decrease transcription of genes in adipocytes and modulate the glucose uptake system [162]. It has been shown to act via the Wnt pathway [163, 164] and also to promote bone growth [164].

In-Utero Exposures

The word obesogen was coined in 2006 by Grun and Bloomberg at the University of California Irvine, who observed that mice exposed to tributyltin (TBT) in-utero became obese regardless of their diet [165]. An obesogen is a chemical compound that promotes obesity by expanding the number of fat cells, increasing the storage of fat in existing fat cells, altering the energy balance to favor calorie storage, or altering the mechanisms by which the body regulates appetite and satiety [166]. The study of these obesogens and the mechanisms by which they act is a rapidly expanding field in endocrinology. One of the first studied obesogens is smoking; maternal smoking is associated with low birth weight and then an increased risk for obesity and metabolic syndrome [167]. Low birthweight is associated with many maternal conditions and is generally correlated with an increased risk of obesity and metabolic syndrome later in life; it is proposed that the fetus that developed under hypoxic or low nutrient conditions cannot adapt to a more nutrient rich environment later in life. Study of these mechanisms is ongoing [168-170]. Maternal overnutrition, weight gains in excess, and a high fat diet have variable reported effects on birthweight, but adult offspring consistently exhibit obesity and metabolic abnormalities [171-173].

Clinical Implications

The clinical implications of the adipogenesis pathway are wide-ranging, and infiltration of adipose tissue is found in many pathologic conditions. Adipose tissue is found to be deposited between ocular muscle fibers in thyroxocytosis [174], it deposits in bone in osteoporosis [175] and it is implicated in the pathogenesis of acquired and congenital lipodystrophies [78] as well as in obesity and the metabolic syndrome.

Obesity and the Metabolic Syndrome

The combination of metabolic abnormalities known as the metabolic syndrome—insulin resistance, hypertension and dyslipidemia have been described since the 1920's. Abdominal obesity was added in the 1980s and the combination has been known as metabolic syndrome or syndrome X ever since [176]. There are several definitions of the metabolic syndrome and its pathophysiology is just beginning to be understood. Although there is still some discrepancy over the definition, studies have estimated a greater than 30% prevalence of metabolic syndrome in Europe and greater than 40% in the United States. Currently the thought is that the preceeding factor is obesity, with the excess adipose tissue creating an inflammatory environment that leads to the other changes. There are differences in risk by sex, race and age. [177]

Over a third of Americans are considered obese and another third are overweight; every week a national newspaper somewhere in the world reports on the prevalence of obesity and the risk it carries for our health [59]. Excess adiposity can occur by tissue hyperplasia (increased cell numbers), or hypertrophy, (increased cell size), with the latter being more associated with insulin resistance [141]. It is associated with leukocyte infiltration and the secretion of pro-inflammatory cytokines, chemokines and adipokines by either the adipocytes themselves or the leukocytes and stromal cells [178]. The mechanisms by which adipocyte hypertrophy induces inflammation and insulin resistance have yet to be established. Hypotheses include hypoxia-induced transcriptional changes, cellular stress from hypoxia, overnutrition (or some combination of the two), and an increase in fatty acids [15]. It has been established that in humans, macrophage infiltration is correlated with both adipocyte size and Body Mass Index (BMI), and it is reduced after surgery-induced weight loss. Macrophage marker expression has also been established as a negative predictor of insulin sensitivity [179]. In mice, obesity induces a change in phenotype of tissue macrophages, from an M2 type that protects against insulin resistance to an M1 inflammatory state that confers sensitivity to insulin resistance; mice fed a high fat diet had higher circulating cells of the M1 type [180]. In humans, increasing BMI is correlated with an increase in M1 macrophages, and the ratio of M1 to M2 cells decreased after weight loss [181]. T Lymphocyte alterations accompany and may even precede macrophage modulation and have been the subject of several recent studies. Cytokines and chemokines produced by adipocytes, leukocytes and stromal cells in adipose tissue exacerbate the inflammatory state. TNF-alpha was the first inflammatory factor found to be secreted from adipose tissue and is considered a primary contributor to adipose dysfunction and obesity [182]. Inflammatory stimuli inhibit PPAR-y function, [63] resulting in impaired adipogenesis, de-differentiation of mature adipocytes, reduced insulin signaling, and increased secretion of adipokines [183].

Predisposition to metabolic syndrome seems to be at least in part governed by genetics. One study showed that adipogenic and lipogenic genes were shown to be more highly expressed in the peripheral tissue of normal weight black women when compared to white women, whereas they were downregulated to a greater extent in obese black women when compared to white women, possibly explaining at least in part, black women's greater susceptibility to type two diabetes [184].

There is also a role for stress hormones and the activation of inflammatory pathways; both catecholamines and steroids have been shown to adversely affect adipose tissue insulin sensitivity and lipid metabolism [185]. Mice lacking in the enzyme that catalyzes the reaction of inactive cortisone to active cortisol were found to be resistant to stress or obesity-induced hyperglycemia [186]. Endotoxin has been shown to cause insulin resistance in humans via the activation of innate immunity [187].

Adipogenesis, Osteogenesis and Hematopoesis

As humans age, there is a tendency to increase the proportion of body weight made up of adipose tissue. There is a concomitant decline in bone marrow density and hematopoiesis [188]. There has been recent focus on the interaction between fat, bone and blood [189]. Literature from the 1800s describes the highly developed capillary systems which formed a network through the "fat glands" [10]. The earliest event associated with adipogenesis is a proliferating network of capillaries in subcutaneous loose connective tissue; there is a spatial and temporal relation with the development of blood vessels and the development of adipose tissue [190]. A later study by the same group showed that this growth may be regulated by transforming growth factor type beta (TGF-beta) [191]. There is a growing body of evidence about the signaling between capillary endothelial cells, extracellular components, and cell-cell interactions. Integrin and plasminogen activator inhibitor guide preadipocyte migration toward developing capillary networks [192]. The inhibition of PPARγ inhibits angiogenesis as well as adipogenesis, and likewise, the blockage of VEGFR-2 inhibits both angiogenesis and adipogenesis [193]. Hypoxia is another important regulator of vascular remodeling, and thus also of adipose tissue remodeling. Like endothelial tissues, adipose tissue produces angiogenic factors in response to hypoxia, which regulate vasculogenesis and maintain preadipocyte viability [194, 195]. Since the 1930s, angiogenic factors have been known to promote healing in ischemic tissue including myocardium [196]. Severe hypoxia has been shown to cause adipocyte death, but preadipocytes remain viable and promote healing [197]. The expression of hypoxia-induced factors has been shown to be different between visceral and subcutaneous fat [198].

Among the first studies to report the inverse relationship between adipocytic and osteogenic cells was done by Beresford et al., They found that when steroid was present through primary and secondary culture, osteocytes predominated, but when steroid was present through secondary culture only, adipocytes predominated [199]. Senescence-accelerated mice were shown to develop osteoporosis within a few months of birth [200] and then it was shown that they not only had decreased osteoblastogenesis, but also increased numbers of adipocytes [201]. Increasing concentrations of dexamethasone resulted in an increase in the number of cells that expressed aP2, a fat-specific gene [202]; they also showed that pluripotent cells turn into adipocytes preferentially when transplanted into steroid-treated mice [203]. Lovastatin was shown to reverse the effects of steroid on bone [204] and the proposed mechanism is by PPARγ2 suppression [205]. Alcohol has also been shown to have a positive effect on adipogenesis and adipocyte hypertrophy and a negative effect on osteogenesis [206]. Homozygous PPARγ deficient stem cells failed to differentiate into adipocytes and spontaneously differentiated into osteoblasts; haploinsufficiency of PPARγ was shown to increase osteoblastogenesis in vitro and bone mass in mice in vivo [207]. PPARγ expression was shown to be elevated in 26-week old mice as compared with eight-week old mice [208]; these stem cells were also shown to differentiate into adipocytes in-vivo without stimulus. Activation of runt-related transcription factor 2 (Runx2) is the initiating event in the commitment to the osteo/chondroprogenitor line; members of the beta catenin family along the Wnt pathway are also involved. Bone morphogenic proteins, osterix, Msh homeobox and hedgehog have also shown to be important; the counterpart to PPARγ, CCAAT/EBPs, glucocorticoid receptors, insulin and KLF5 [209]. Several mutations along the Wnt pathway have also been linked to human disease; loss of lipoprotein-related protein 5 (LRP5), a Wnt receptor, causes an osteoporosis-pseudoglioma syndrome characterized by reduced bone mass and defects in osteoblast differentiation [210].

Thoughts on Further Research

In the nearly four decades since the first line of preadipocytes were cultured, a huge body of research has amassed, and much has been learned about the differentiation of adipocytes. There is still much to be learned about precisely how this process works and how this work can be translated to therapeutics. The newest area of research--the interaction between osteogenesis, adipogenesis and hematopoiesis, presents unique challenges to researchers. Although there have been significant research discoveries made using cultured cell lines, and we have come to understand those cells very well, we know that adipogenesis in vivo is influenced by factors that are more difficult to regulate, like cell matrix factors and environmental toxins. There is so far no standardized protocol for isolation of mesenchymal stem cells; these isolates comprise many cell types and this impacts upon the differentiation potential of single clones. The biggest therapeutic discoveries, thiazolidinediones, which have been used for diabetes treatment, are also being tested for a variety of other therapeutic implications. Recently there have been studies linking thiazolidinediones to a possible increased risk of myocardial infarction and death [211, 212]; several prostaglandins have been implicated as potential endogenous PPARγ ligands [213]. The search for endogenous PPARγ ligands and the search for synthetic ligands that more closely mimic endogenous ligands will continue to be an area of interest for researchers.

There is also promise in treating the inflammation that is associated with obesity. Treatment of rheumatoid arthritis patients with infliximab, an anti-TNF-alpha agent, resulted in reduction of serum insulin levels [214]; it has yet to be determined whether there is a role for infliximab in patients without inflammatory disease. Inflammatory kinases could also be targets for therapeutics. Salsalate, a precursor of salicylic acid, has been used successfully to improve glycemic control in diabetics without adverse outcomes [215].

Despite considerable research effort, there is still no effective drug to counteract weight gain or reverse the metabolic syndrome. Brown fat, described as "neither fat nor flesh" by Swiss scientist Konrad Gessner in the 1500's, is going to be a target of future research efforts. Brown fat regulates body temperature by dissipating heat through non-shivering thermogenesis via the uncoupling of the electron transport chain. For years, brown fat was assumed to be absent in adults, or so scarce that it could not make a significant contribution to energy expenditure [216]. Recently PET/CT scans have demonstrated the existence of thermogenic adipose tissue in adults. Early studies have indicated a relationship between brown adipocytes and muscle cells [217, 218]. White fat cells have been shown to develop UCP1-positive islets within white adipose deposits following cold or beta-adrenergic stimulation. Mice that have downregulation of VEGF have increased resistance to diet-induced obesity and had increased brown fat [219]. There much promise to brown fat, but still much that is to be discovered. In vivo discrepancies have been found between mouse and human brown adipose tissue amount, location and activity. Few human models exist, and in contrast to white adipose tissue, which is readily accessible, brown adipose tissue, found surrounding the great vessels, around the kidneys, and around the spine, is relatively difficult to access. So far only one human derived cell line is described in the literature [220] and so far no drugs or therapies have been developed that increase the amount of brown fat.

Conclusion

Adipogenesis is perhaps the best studied field in cellular biology, but there are many questions yet to be answered about the process and how it relates to human health. In further understanding this pathway we will also further our understanding of the factors that influence mesenchymal cell fate to osteogenic, chondrogenic and other cell lines. Our increasing understanding of the metabolic intricacies of fat has paralleled the increasing prevalence of obesity and related diseases in the developed world. It will be incumbent on researchers and clinicians alike to accelerate our understanding of adipogenesis, not only to combat the emerging obesity epidemic, but also harness the untapped potential of lipid metabolism pathways for other human conditions such as aging, wound and fracture healing, insulin resistance, cancer and a myriad of other applications.

The ability to store fat conveyed survival advantages to humans for thousands of years. For the vast majority of human history and pre-history, the daily struggle for food and survival made corpulence a mere curiosity. Famine and starvation were widespread, and while the risks associated with obesity were observed as early as Hippocrates' time, these risks were not universal, nor are the risks immediate, whereas everyone who goes without food for long enough will die, slowly and painfully, usually with their loved ones, as famine rarely affects individuals in isolation. We must find a way to overcome what seems to be a predisposition toward creating fat cells at the expense of bone, muscle tissue and cartilage. Ironically, our success as a species has created an obesity problem; we must expand our knowledge of adipogenesis and outpace our lipid excess with translatable lipid research to ensure a healthy future for the coming generations.

References

[1] Oxford University Press., The Holy Bible : containing the Old and New Testaments: New Revised Standard Version. Anglicized ed. 1995, Oxford; New York: Oxford University Press. xviii, 905, 267 p.
[2] Roth, J., Evolutionary speculation about tuberculosis and the metabolic and inflammatory processes of obesity. *JAMA*, 2009. 301(24): p. 2586-8.
[3] Vague, J., [Not Available]. *Presse Med.*, 1947. 55(30): p. 339.
[4] Hippocrates, L. Verhoofd, and E. Marks, The aphorisms of Hippocrates. 1818, New York,: Collins & co.
[5] Cannon, B. and J. Nedergaard, Developmental biology: Neither fat nor flesh. *Nature*, 2008. 454(7207): p. 947-8.
[6] Shaw, H. B., A Contribution to the Study of the Morphology of Adipose Tissue. *J. Anat. Physiol.*, 1901. 36(Pt 1): p. 1-13.
[7] Kolliker, W., Verh Bd 7, 1886. p. 183.
[8] Flemming, *Achiv fur mikroskopiche Anatomie*, 1871. 7: p. 321.
[9] Czajkawitz, Reichert u du Bois Reymond. *Archiv*. 1866.
[10] Toldt, Sitzungsberichte de Akademie der Wisseschaften. Sitzungsberichte de Akademie der Wisseschaften, 1870. 62: p. 445.
[11] Waldeyer, Archiv fur mikroskopische. *Anatomie Band*, 1875. 11: p. 176.
[12] Sypniewska, G., Regulation of new fat cell formation. *Acta Physiol. Pol.*, 1989. 40(2): p. 156-63.
[13] Siiteri, P. K., Adipose tissue as a source of hormones. *Am. J. Clin. Nutr.*, 1987. 45(1 Suppl): p. 277-82.
[14] Zhang, Y., et al., Positional cloning of the mouse obese gene and its human homologue. *Nature*, 1994. 372(6505): p. 425-32.
[15] Wellen, K. E. and G. S. Hotamisligil, Obesity-induced inflammatory changes in adipose tissue. *J. Clin. Invest.*, 2003. 112(12): p. 1785-1788.
[16] Badman, M. K. and J. S. Flier, The adipocyte as an active participant in energy balance and metabolism. *Gastroenterology*, 2007. 132(6): p. 2103-2115.
[17] Galic, S., J. S. Oakhill, and G. R. Steinberg, Adipose tissue as an endocrine organ. *Mol. Cell Endocrinol.*, 2010. 316(2): p. 129-139.
[18] Caplan, A. I. and S. P. Bruder, Mesenchymal stem cells: building blocks for molecular medicine in the 21st century. *Trends Mol. Med.*, 2001. 7(6): p. 259-64.
[19] Enerback, S., The origins of brown adipose tissue. *N. Engl. J. Med.*, 2009. 360(19): p. 2021-2023.

[20] Cinti, S., The adipose organ: morphological perspectives of adipose tissues. *Proc. Nutr. Soc.*, 2001. 60(3): p. 319-28.
[21] Gesta, S., Y.-H. Tseng, and C.R. Kahn, Developmental origin of fat: tracking obesity to its source. *Cell*, 2007. 131(2): p. 242-256.
[22] Aquila, H., T. A. Link, and M. Klingenberg, The uncoupling protein from brown fat mitochondria is related to the mitochondrial ADP/ATP carrier. Analysis of sequence homologies and of folding of the protein in the membrane. *EMBO J.*, 1985. 4(9): p. 2369-76.
[23] Sethi, J. V.-P. A., Adipokines in Health and Disease, in Metabolic basis of obesity, R.S. Ahima, Editor 2011, Springer: New York. p. xiv, 391 p.2.
[24] Taylor, S. M. and P. A. Jones, Changes in phenotypic expression in embryonic and adult cells treated with 5-azacytidine. *J. Cell Physiol.*, 1982. 111(2): p. 187-194.
[25] Green, H. and O. Kehinde, An established preadipose cell line and its differentiation in culture. II. Factors affecting the adipose conversion. *Cell*, 1975. 5(1): p. 19-27.
[26] Rosen, E. D. and B. M. Spiegelman, Molecular regulation of adipogenesis. *Annu. Rev. Cell Dev. Biol.*, 2000. 16: p. 145-171.
[27] Scroyen, I., B. Hemmeryckx, and H. R. Lijnen, From mice to men: mouse models in obesity research: What can we learn? *Thromb. Haemost.*, 2013. 109(5).
[28] Lanotte, M., et al., Clonal preadipocyte cell lines with different phenotypes derived from murine marrow stroma: factors influencing growth and adipogenesis in vitro. *J. Cell Physiol.*, 1982. 111(2): p. 177-186.
[29] Modica, S. and C. Wolfrum, Bone morphogenic proteins signaling in adipogenesis and energy homeostasis. *Biochim. Biophys. Acta*, 2013. 1831(5): p. 915-923.
[30] Du, B., et al., The transcription factor paired-related homeobox 1 (Prrx1) inhibits adipogenesis by activating transforming growth factor-β (TGFβ) signaling. *J. Biol. Chem.*, 2013. 288(5): p. 3036-3047.
[31] Gregoire, F. M., C. M. Smas, and H. S. Sul, Understanding adipocyte differentiation. *Physiol. Rev.*, 1998. 78(3): p. 783-809.
[32] Pairault, J. and H. Green, A study of the adipose conversion of suspended 3T3 cells by using glycerophosphate dehydrogenase as differentiation marker. *Proc. Natl. Acad. Sci. USA*, 1979. 76(10): p. 5138-5142.

[33] Gregoire, F., et al., Interferon-gamma and interleukin-1 beta inhibit adipoconversion in cultured rodent preadipocytes. *J. Cell Physiol.*, 1992. 151(2): p. 300-9.

[34] Chapman, A. B., et al., Analysis of gene expression during differentiation of adipogenic cells in culture and hormonal control of the developmental program. *J. Biol. Chem.*, 1984. 259(24): p. 15548-55.

[35] Harrison, J. J., E. Soudry, and R. Sager, Adipocyte conversion of CHEF cells in serum-free medium. *J. Cell Biol.*, 1985. 100(2): p. 429-434.

[36] Chapman, A. B., D. M. Knight, and G. M. Ringold, Glucocorticoid regulation of adipocyte differentiation: hormonal triggering of the developmental program and induction of a differentiation-dependent gene. *J. Cell Biol.*, 1985. 101(4): p. 1227-35.

[37] Ringold, G. M., et al., Hormonal control of adipogenesis. *Ann. N. Y. Acad. Sci.*, 1986. 478: p. 109-119.

[38] Yang, W., et al., Regulation of adipogenesis by cytoskeleton remodelling is facilitated by acetyltransferase MEC-17-dependent acetylation of alpha-tubulin. *Biochem. J.*, 2013. 449(3): p. 605-12.

[39] Chang, H. and M. L. Knothe Tate, Structure-function relationships in the stem cell's mechanical world B: emergent anisotropy of the cytoskeleton correlates to volume and shape changing stress exposure. *Mol. Cell Biomech.*, 2011. 8(4): p. 297-318.

[40] Titushkin, I., et al., Control of adipogenesis by ezrin, radixin and moesin-dependent biomechanics remodeling. *J. Biomech.*, 2013. 46(3): p. 521-6.

[41] Brekken, R. A. and E. H. Sage, SPARC, a matricellular protein: at the crossroads of cell-matrix communication. *Matrix Biol.*, 2001. 19(8): p. 816-27.

[42] Bradshaw, A. D., et al., SPARC-null mice exhibit increased adiposity without significant differences in overall body weight. *Proc. Natl. Acad. Sci. USA*, 2003. 100(10): p. 6045-50.

[43] Delany, A. M., et al., Osteonectin-null mutation compromises osteoblast formation, maturation, and survival. *Endocrinology*, 2003. 144(6): p. 2588-96.

[44] Shao, D. and M. A. Lazar, Peroxisome proliferator activated receptor gamma, CCAAT/enhancer-binding protein alpha, and cell cycle status regulate the commitment to adipocyte differentiation. *J. Biol. Chem.*, 1997. 272(34): p. 21473-21478.

[45] Morrison, R. F. and S. R. Farmer, Role of PPARgamma in regulating a cascade expression of cyclin-dependent kinase inhibitors, p18(INK4c)

and p21(Waf1/Cip1), during adipogenesis. *J. Biol. Chem.*, 1999. 274(24): p. 17088-97.

[46] Altiok, S., M. Xu, and B. M. Spiegelman, PPARgamma induces cell cycle withdrawal: inhibition of E2F/DP DNA-binding activity via down-regulation of PP2A. *Genes Dev.*, 1997. 11(15): p. 1987-98.

[47] Entenmann, G. and H. Hauner, Relationship between replication and differentiation in cultured human adipocyte precursor cells. *Am. J. Physiol.*, 1996. 270(4 Pt 1): p. C1011-6.

[48] Darlington, G. J., S. E. Ross, and O. A. MacDougald, The role of C/EBP genes in adipocyte differentiation. *J. Biol. Chem.*, 1998. 273(46): p. 30057-60.

[49] Spiegelman, B. M., et al., Regulation of adipocyte gene expression in differentiation and syndromes of obesity/diabetes. *J. Biol. Chem.*, 1993. 268(10): p. 6823-6.

[50] Mandrup, S. and M. D. Lane, Regulating adipogenesis. *J. Biol. Chem.*, 1997. 272(9): p. 5367-70.

[51] Meissburger, B., et al., Tissue inhibitor of matrix metalloproteinase 1 (TIMP1) controls adipogenesis in obesity in mice and in humans. *Diabetologia*, 2011. 54(6): p. 1468-79.

[52] Hausman, G. J., Meat Science and Muscle Biology Symposium: the influence of extracellular matrix on intramuscular and extramuscular adipogenesis. *J. Anim. Sci.*, 2012. 90(3): p. 942-9.

[53] Adams, M., et al., Transcriptional activation by peroxisome proliferator-activated receptor gamma is inhibited by phosphorylation at a consensus mitogen-activated protein kinase site. *J. Biol. Chem.*, 1997. 272(8): p. 5128-32.

[54] Djian, P., M. Phillips, and H. Green, The activation of specific gene transcription in the adipose conversion of 3T3 cells. *J. Cell Physiol.*, 1985. 124(3): p. 554-6.

[55] Lefebvre, A. M., et al., Depot-specific differences in adipose tissue gene expression in lean and obese subjects. *Diabetes*, 1998. 47(1): p. 98-103.

[56] Morimoto, C., T. Tsujita, and H. Okuda, Norepinephrine-induced lipolysis in rat fat cells from visceral and subcutaneous sites: role of hormone-sensitive lipase and lipid droplets. *J. Lipid Res.*, 1997. 38(1): p. 132-8.

[57] Ostman, J., et al., Regional differences in the control of lipolysis in human adipose tissue. *Metabolism*, 1979. 28(12): p. 1198-205.

[58] Wahrenberg, H., F. Lonnqvist, and P. Arner, Mechanisms underlying regional differences in lipolysis in human adipose tissue. *J. Clin. Invest.*, 1989. 84(2): p. 458-67.

[59] Tchernof, A. and J. P. Despres, Pathophysiology of human visceral obesity: an update. *Physiol. Rev.*, 2013. 93(1): p. 359-404.

[60] Ranadive, S. A. and C. Vaisse, Lessons from extreme human obesity: monogenic disorders. *Endocrinol. Metab. Clin. North Am.*, 2008. 37(3): p. 733-51, x.

[61] Sandholt, C. H., T. Hansen, and O. Pedersen, Beyond the fourth wave of genome-wide obesity association studies. *Nutr. Diabetes*, 2012. 2.

[62] Lowe, C. E., S. O'Rahilly, and J.J. Rochford, Adipogenesis at a glance. *J. Cell Sci.*, 2011. 124(Pt 16): p. 2681-6.

[63] Lehrke, M. and M. A. Lazar, The many faces of PPARgamma. *Cell*, 2005. 123(6): p. 993-999.

[64] Graves, R. A., P. Tontonoz, and B.M. Spiegelman, Analysis of a tissue-specific enhancer: ARF6 regulates adipogenic gene expression. *Mol. Cell Biol.*, 1992. 12(7): p. 3313-3313.

[65] Tontonoz, P., et al., mPPAR gamma 2: tissue-specific regulator of an adipocyte enhancer. *Genes Dev.*, 1994. 8(10): p. 1224-34.

[66] Braissant, O., et al., Differential expression of peroxisome proliferator-activated receptors (PPARs): tissue distribution of PPAR-alpha, -beta, and -gamma in the adult rat. *Endocrinology*, 1996. 137(1): p. 354-66.

[67] Chandra, V., et al., Structure of the intact PPAR-gamma-RXR- nuclear receptor complex on DNA. *Nature*, 2008. 456(7220): p. 350-6.

[68] Chawla, A., et al., Nuclear receptors and lipid physiology: opening the X-files. *Science*, 2001. 294(5548): p. 1866-70.

[69] Barak, Y., et al., PPAR gamma is required for placental, cardiac, and adipose tissue development. *Mol. Cell*, 1999. 4(4): p. 585-95.

[70] Rosen, E. D., et al., PPAR gamma is required for the differentiation of adipose tissue in vivo and in vitro. *Mol. Cell*, 1999. 4(4): p. 611-7.

[71] Freedman, B. D., et al., A dominant negative peroxisome proliferator-activated receptor-gamma knock-in mouse exhibits features of the metabolic syndrome. *J. Biol. Chem.*, 2005. 280(17): p. 17118-25.

[72] Agostini, M., et al., Non-DNA binding, dominant-negative, human PPARgamma mutations cause lipodystrophic insulin resistance. *Cell Metab.*, 2006. 4(4): p. 303-11.

[73] Tsai, Y. S. and N. Maeda, PPARgamma: a critical determinant of body fat distribution in humans and mice. *Trends Cardiovasc. Med.*, 2005. 15(3): p. 81-5.

[74] Ristow, M., et al., Obesity associated with a mutation in a genetic regulator of adipocyte differentiation. *N. Engl. J. Med.*, 1998. 339(14): p. 953-9.

[75] Beamer, B. A., et al., Association of the Pro12Ala variant in the peroxisome proliferator-activated receptor-gamma2 gene with obesity in two Caucasian populations. *Diabetes*, 1998. 47(11): p. 1806-8.

[76] Hara, K., et al., The role of PPARgamma as a thrifty gene both in mice and humans. *Br. J. Nutr.*, 2000. 84 Suppl 2: p. S235-9.

[77] Yen, C. J., et al., Molecular scanning of the human peroxisome proliferator activated receptor gamma (hPPAR gamma) gene in diabetic Caucasians: identification of a Pro12Ala PPAR gamma 2 missense mutation. *Biochem. Biophys. Res. Commun.*, 1997. 241(2): p. 270-4.

[78] Auclair, M., et al., Peroxisome proliferator-activated receptor-gamma mutations responsible for lipodystrophy with severe hypertension activate the cellular renin-angiotensin system. *Arterioscler. Thromb. Vasc. Biol.*, 2013. 33(4): p. 829-38.

[79] Yki-Jarvinen, H., Thiazolidinediones. *N. Engl. J. Med.*, 2004. 351(11): p. 1106-18.

[80] Sohda, T., et al., Studies on antidiabetic agents. I. Synthesis of 5-[4-(2-methyl-2-phenylpropoxy)-benzyl]thiazolidine-2,4-dione (AL-321) and related compounds. *Chem. Pharm. Bull.* (Tokyo), 1982. 30(10): p. 3563-73.

[81] Fujita, T., et al., Reduction of insulin resistance in obese and/or diabetic animals by 5-[4-(1-methylcyclohexylmethoxy)benzyl]-thiazolidine-2,4-dione (ADD-3878, U-63,287, ciglitazone), a new antidiabetic agent. *Diabetes*, 1983. 32(9): p. 804-810.

[82] Kobayashi, M., et al., A new potentiator of insulin action. Post-receptor activation in vitro. *FEBS Lett.*, 1983. 163(1): p. 50-53.

[83] Tontonoz, P., et al., Terminal differentiation of human liposarcoma cells induced by ligands for peroxisome proliferator-activated receptor gamma and the retinoid X receptor. *Proc. Natl. Acad. Sci. USA*, 1997. 94(1): p. 237-41.

[84] Jiang, C., A. T. Ting, and B. Seed, PPAR-gamma agonists inhibit production of monocyte inflammatory cytokines. *Nature*, 1998. 391 (6662): p. 82-86.

[85] Ricote, M., et al., The peroxisome proliferator-activated receptor (PPARgamma) as a regulator of monocyte/macrophage function. *J. Leukoc. Biol.*, 1999. 66(5): p. 733-9.

[86] Igarashi, M., et al., Pioglitazone reduces atherogenic outcomes in type 2 diabetic patients. *J. Atheroscler. Thromb.*, 2008. 15(1): p. 34-40.
[87] Zeender, E., et al., Pioglitazone and sodium salicylate protect human beta-cells against apoptosis and impaired function induced by glucose and interleukin-1beta. *J. Clin. Endocrinol. Metab.*, 2004. 89(10): p. 5059-66.
[88] Schmidt, S., et al., Anti-inflammatory and antiproliferative actions of PPAR-gamma agonists on T lymphocytes derived from MS patients. *J. Leukoc. Biol.*, 2004. 75(3): p. 478-85.
[89] Malhotra, A., et al., Thiazolidinediones for plaque psoriasis: a systematic review and meta-analysis. *Evid. Based Med.*, 2012. 17(6): p. 171-6.
[90] Zhang, Q., et al., PPARgamma agonist rosiglitazone is neuroprotective after traumatic spinal cord injury via anti-inflammatory in adult rats. *Neurol. Res.*, 2010. 32(8): p. 852-9.
[91] Shimada, T. and H. Hiraishi, [PPARgamma agonists for the diseases of gastrointestinal tract and liver]. *Nihon Rinsho*, 2010. 68(2): p. 312-6.
[92] Lin, T. H., et al., 15-deoxy-Delta(12,14) -prostaglandin-J2 and ciglitazone inhibit TNF-alpha-induced matrix metalloproteinase 13 production via the antagonism of NF-kappaB activation in human synovial fibroblasts. *J. Cell Physiol.*, 2011. 226(12): p. 3242-50.
[93] Forman, B. M., J. Chen, and R. M. Evans, Hypolipidemic drugs, polyunsaturated fatty acids, and eicosanoids are ligands for peroxisome proliferator-activated receptors alpha and delta. *Proc. Natl. Acad. Sci. USA*, 1997. 94(9): p. 4312-7.
[94] Kliewer, S. A., et al., Fatty acids and eicosanoids regulate gene expression through direct interactions with peroxisome proliferator-activated receptors alpha and gamma. *Proc. Natl. Acad. Sci. USA*, 1997. 94(9): p. 4318-23.
[95] Dubois, S. G., et al., Potential role of increased matrix metalloproteinase-2 (MMP2) transcription in impaired adipogenesis in type 2 diabetes mellitus. *Biochem. Biophys. Res. Commun.*, 2008. 367(4): p. 725-8.
[96] Lehmann, J. M., et al., An antidiabetic thiazolidinedione is a high affinity ligand for peroxisome proliferator-activated receptor gamma (PPAR gamma). *J. Biol. Chem.*, 1995. 270(22): p. 12953-12956.
[97] Lefterova, M. I., et al., PPARgamma and C/EBP factors orchestrate adipocyte biology via adjacent binding on a genome-wide scale. *Genes Dev.*, 2008. 22(21): p. 2941-52.

[98] Nielsen, R., et al., Genome-wide profiling of PPARgamma:RXR and RNA polymerase II occupancy reveals temporal activation of distinct metabolic pathways and changes in RXR dimer composition during adipogenesis. *Genes Dev*., 2008. 22(21): p. 2953-67.

[99] Louet, J. F. and B. W. O'Malley, Coregulators in adipogenesis: what could we learn from the SRC (p160) coactivator family? *Cell Cycle*, 2007. 6(20): p. 2448-52.

[100] Guan, H.-P., et al., Corepressors selectively control the transcriptional activity of PPARgamma in adipocytes. Genes Dev., 2005. 19(4): p. 453-461.

[101] Lekstrom-Himes, J. and K. G. Xanthopoulos, Biological role of the CCAAT/enhancer-binding protein family of transcription factors. *J. Biol. Chem*., 1998. 273(44): p. 28545-8.

[102] Otto, T. C. and M. D. Lane, Adipose development: from stem cell to adipocyte. *Crit. Rev. Biochem. Mol. Biol*., 2005. 40(4): p. 229-42.

[103] Cao, Z., R. M. Umek, and S. L. McKnight, Regulated expression of three C/EBP isoforms during adipose conversion of 3T3-L1 cells. *Genes Dev*., 1991. 5(9): p. 1538-52.

[104] Wang, N. D., et al., Impaired energy homeostasis in C/EBP alpha knockout mice. *Science*, 1995. 269(5227): p. 1108-12.

[105] Linhart, H. G., et al., C/EBPalpha is required for differentiation of white, but not brown, adipose tissue. *Proc. Natl. Acad. Sci. USA*, 2001. 98(22): p. 12532-7.

[106] Tanaka, T., et al., Defective adipocyte differentiation in mice lacking the C/EBPbeta and/or C/EBPdelta gene. *EMBO J*., 1997. 16(24): p. 7432-43.

[107] Wu, Z., N. L. Bucher, and S. R. Farmer, Induction of peroxisome proliferator-activated receptor gamma during the conversion of 3T3 fibroblasts into adipocytes is mediated by C/EBPbeta, C/EBPdelta, and glucocorticoids. *Mol. Cell Biol*., 1996. 16(8): p. 4128-36.

[108] Ross, S. E., et al., Inhibition of adipogenesis by Wnt signaling. *Science*, 2000. 289(5481): p. 950-3.

[109] Arango, N. A., et al., Conditional deletion of beta-catenin in the mesenchyme of the developing mouse uterus results in a switch to adipogenesis in the myometrium. *Dev. Biol*., 2005. 288(1): p. 276-83.

[110] Bennett, C. N., et al., Regulation of osteoblastogenesis and bone mass by Wnt10b. *Proc. Natl. Acad. Sci. USA*, 2005. 102(9): p. 3324-9.

[111] Qin, L., et al., A deep investigation into the adipogenesis mechanism: profile of microRNAs regulating adipogenesis by modulating the

canonical Wnt/beta-catenin signaling pathway. *BMC Genomics*, 2010. 11: p. 320.

[112] Birsoy, K., Z. Chen, and J. Friedman, Transcriptional regulation of adipogenesis by KLF4. *Cell Metab.*, 2008. 7(4): p. 339-47.

[113] Jack, B. H. and M. Crossley, GATA proteins work together with friend of GATA (FOG) and C-terminal binding protein (CTBP) co-regulators to control adipogenesis. *J. Biol. Chem.*, 2010. 285(42): p. 32405-14.

[114] Xu, Y., et al., Downregulation of GATA-2 and overexpression of adipogenic gene-PPARgamma in mesenchymal stem cells from patients with aplastic anemia. *Exp. Hematol.*, 2009. 37(12): p. 1393-9.

[115] Chen, L., et al., Overexpression of C10orf116 promotes proliferation, inhibits apoptosis and enhances glucose transport in 3T3-L1 adipocytes. *Mol. Med. Rep.*, 2013. 7(5): p. 1477-81.

[116] Shimba, S., et al., Brain and muscle Arnt-like protein-1 (BMAL1), a component of the molecular clock, regulates adipogenesis. *Proc. Natl. Acad. Sci. USA*, 2005. 102(34): p. 12071-6.

[117] Wang, L. M., et al., Role for the NR2B subunit of the N-methyl-D-aspartate receptor in mediating light input to the circadian system. *Eur. J. Neurosci.*, 2008. 27(7): p. 1771-9.

[118] Le Magnen, J., [Lipogenesis, lipolysis and feeding rhythms]. *Ann. Endocrinol.* (Paris), 1988. 49(2): p. 98-104.

[119] Ando, H., et al., Rhythmic messenger ribonucleic acid expression of clock genes and adipocytokines in mouse visceral adipose tissue. *Endocrinology*, 2005. 146(12): p. 5631-6.

[120] Gonzalez, A., et al., Melatonin promotes differentiation of 3T3-L1 fibroblasts. *J. Pineal Res.*, 2012. 52(1): p. 12-20.

[121] Musri, M. M., R. Gomis, and M. Parrizas, A chromatin perspective of adipogenesis. *Organogenesis*, 2010. 6(1): p. 15-23.

[122] Musri, M. M., et al., Histone demethylase LSD1 regulates adipogenesis. *J. Biol. Chem.*, 2010. 285(39): p. 30034-41.

[123] Musri, M. M. and M. Parrizas, Epigenetic regulation of adipogenesis. *Curr. Opin. Clin. Nutr. Metab. Care*, 2012. 15(4): p. 342-9.

[124] Siersbaek, R., R. Nielsen, and S. Mandrup, Transcriptional networks and chromatin remodeling controlling adipogenesis. *Trends Endocrinol. Metab.*, 2012. 23(2): p. 56-64.

[125] Steger, D. J., et al., Propagation of adipogenic signals through an epigenomic transition state. *Genes Dev.*, 2010. 24(10): p. 1035-44.

[126] Pessler-Cohen, D., et al., GLUT4 repression in response to oxidative stress is associated with reciprocal alterations in C/EBP alpha and delta

isoforms in 3T3-L1 adipocytes. *Arch. Physiol. Biochem.*, 2006. 112(1): p. 3-12.

[127] Minet, E., et al., Transduction pathways involved in Hypoxia-Inducible Factor-1 phosphorylation and activation. *Free Radic. Biol. Med.*, 2001. 31(7): p. 847-55.

[128] Richard, D. E., E. Berra, and J. Pouyssegur, Angiogenesis: how a tumor adapts to hypoxia. *Biochem. Biophys. Res. Commun.*, 1999. 266(3): p. 718-22.

[129] Yun, Z., et al., Inhibition of PPAR gamma 2 gene expression by the HIF-1-regulated gene DEC1/Stra13: a mechanism for regulation of adipogenesis by hypoxia. *Dev. Cell*, 2002. 2(3): p. 331-41.

[130] Carriere, A., et al., Mitochondrial reactive oxygen species control the transcription factor CHOP-10/GADD153 and adipocyte differentiation: a mechanism for hypoxia-dependent effect. *J. Biol. Chem.*, 2004. 279(39): p. 40462-9.

[131] Pasarica, M., et al., Reduced oxygenation in human obese adipose tissue is associated with impaired insulin suppression of lipolysis. *J. Clin. Endocrinol. Metab.*, 2010. 95(8): p. 4052-4055.

[132] Pasarica, M., et al., Reduced adipose tissue oxygenation in human obesity: evidence for rarefaction, macrophage chemotaxis, and inflammation without an angiogenic response. *Diabetes*, 2009. 58(3): p. 718-725.

[133] Halberg, N., et al., Hypoxia-inducible factor 1alpha induces fibrosis and insulin resistance in white adipose tissue. *Mol. Cell Biol.*, 2009. 29(16): p. 4467-83.

[134] Rudich, A., et al., Oxidant stress reduces insulin responsiveness in 3T3-L1 adipocytes. *Am. J. Physiol.*, 1997. 272(5 Pt 1): p. E935-40.

[135] Galinier, A., et al., Adipose tissue proadipogenic redox changes in obesity. *J. Biol. Chem.*, 2006. 281(18): p. 12682-7.

[136] Reitman, M. L., Metabolic lessons from genetically lean mice. *Annu. Rev. Nutr.*, 2002. 22: p. 459-82.

[137] Fujita, K., et al., Systemic oxidative stress is associated with visceral fat accumulation and the metabolic syndrome. *Circ. J.*, 2006. 70(11): p. 1437-42.

[138] Matsuzawa-Nagata, N., et al., Increased oxidative stress precedes the onset of high-fat diet-induced insulin resistance and obesity. *Metabolism*, 2008. 57(8): p. 1071-7.

[139] Fisher, G. J., et al., Collagen fragmentation promotes oxidative stress and elevates matrix metalloproteinase-1 in fibroblasts in aged human skin. *Am. J. Pathol.*, 2009. 174(1): p. 101-14.

[140] Furukawa, S., et al., Increased oxidative stress in obesity and its impact on metabolic syndrome. *J. Clin. Invest.*, 2004. 114(12): p. 1752-61.

[141] Gustafson, B., et al., Inflammation and impaired adipogenesis in hypertrophic obesity in man. *Am. J. Physiol. Endocrinol. Metab.*, 2009. 297(5): p. 999-999.

[142] Torti, F. M., et al., Modulation of adipocyte differentiation by tumor necrosis factor and transforming growth factor beta. *J. Cell Biol.*, 1989. 108(3): p. 1105-13.

[143] Vassaux, G., et al., Differential response of preadipocytes and adipocytes to prostacyclin and prostaglandin E2: physiological implications. *Endocrinology*, 1992. 131(5): p. 2393-8.

[144] Chartoumpekis, D. V., et al., Nrf2 activation diminishes during adipocyte differentiation of ST2 cells. *Int. J. Mol. Med.*, 2011. 28(5): p. 823-8.

[145] Xue, P., et al., Adipose deficiency of Nrf2 in ob/ob mice results in severe metabolic syndrome. *Diabetes*, 2013. 62(3): p. 845-854.

[146] Lu, H., W. Cui, and C. D. Klaassen, Nrf2 protects against 2,3,7,8-tetrachlorodibenzo-p-dioxin (TCDD)-induced oxidative injury and steatohepatitis. *Toxicol. Appl. Pharmacol.*, 2011. 256(2): p. 122-35.

[147] Dasuri, K., et al., Proteasome alterations during adipose differentiation and aging: links to impaired adipocyte differentiation and development of oxidative stress. *Free Radic. Biol. Med.*, 2011. 51(9): p. 1727-1735.

[148] Kim, H.-S., et al., Oleanolic acid suppresses resistin induction in adipocytes by modulating Tyk-STAT signaling. *Nutr. Res.*, 2013. 33(2): p. 144-153.

[149] Moghe, S. S., et al., Effect of blueberry polyphenols on 3T3-F442A preadipocyte differentiation. *J. Med. Food*, 2012. 15(5): p. 448-52.

[150] Jungbauer, A. and S. Medjakovic, Phytoestrogens and the metabolic syndrome. *J. Steroid Biochem. Mol. Biol.*, 2013.

[151] Ohnishi, R., et al., 1'-acetoxychavicol acetate inhibits adipogenesis in 3T3-L1 adipocytes and in high fat-fed rats. *Am. J. Chin. Med.*, 2012. 40(6): p. 1189-1204.

[152] Ju, J. H., et al., Anti-obesity and antioxidative effects of purple sweet potato extract in 3T3-L1 adipocytes in vitro. *J. Med. Food*, 2011. 14(10): p. 1097-106.

[153] Lee, Y.-J., et al., Buckwheat (Fagopyrum esculentum M.) Sprout Treated with Methyl Jasmonate (MeJA) Improved Anti-Adipogenic Activity Associated with the Oxidative Stress System in 3T3-L1 Adipocytes. *Int. J. Mol. Sci.*, 2013. 14(1): p. 1428-1442.

[154] Kim, D. and G. D. Han, Ameliorating effects of fermented rice bran extract on oxidative stress induced by high glucose and hydrogen peroxide in 3T3-L1 adipocytes. *Plant Foods Hum. Nutr.*, 2011. 66(3): p. 285-290.

[155] Liang, C.-H., et al., Brazilein from Caesalpinia sappan L. Antioxidant Inhibits Adipocyte Differentiation and Induces Apoptosis through Caspase-3 Activity and Anthelmintic Activities against Hymenolepis nana and Anisakis simplex. *Evid. Based Complement Alternat. Med.*, 2013. 2013: p. 864892-864892.

[156] Kim, H., et al., HPMC supplementation reduces abdominal fat content, intestinal permeability, inflammation, and insulin resistance in diet-induced obese mice. *Mol. Nutr. Food Res.*, 2012. 56(9): p. 1464-1476.

[157] Kim, S., et al., Resveratrol exerts anti-obesity effects via mechanisms involving down-regulation of adipogenic and inflammatory processes in mice. *Biochem. Pharmacol.*, 2011. 81(11): p. 1343-1351.

[158] Rayalam, S., M. A. Della-Fera, and C. A. Baile, Synergism between resveratrol and other phytochemicals: implications for obesity and osteoporosis. *Mol. Nutr. Food Res.*, 2011. 55(8): p. 1177-1185.

[159] Tseng, P. C., et al., Resveratrol promotes osteogenesis of human mesenchymal stem cells by upregulating RUNX2 gene expression via the SIRT1/FOXO3A axis. *J. Bone Miner. Res.*, 2011. 26(10): p. 2552-63.

[160] Baile, C. A., et al., Effect of resveratrol on fat mobilization. *Ann. N. Y. Acad. Sci.*, 2011. 1215: p. 40-7.

[161] Lin, J., M. A. Della-Fera, and C.A. Baile, Green tea polyphenol epigallocatechin gallate inhibits adipogenesis and induces apoptosis in 3T3-L1 adipocytes. *Obes. Res.*, 2005. 13(6): p. 982-90.

[162] Ashida, H., et al., Anti-obesity actions of green tea: possible involvements in modulation of the glucose uptake system and suppression of the adipogenesis-related transcription factors. *Biofactors*, 2004. 22(1-4): p. 135-40.

[163] Freise, C., et al., (+)-Episesamin inhibits adipogenesis and exerts anti-inflammatory effects in 3T3-L1 (pre)adipocytes by sustained Wnt signaling, down-regulation of PPARγ and induction of iNOS. *J. Nutr. Biochem.*, 2013. 24(3): p. 550-555.

[164] Ko, C. H., et al., Pro-bone and antifat effects of green tea and its polyphenol, epigallocatechin, in rat mesenchymal stem cells in vitro. *J. Agric. Food Chem.*, 2011. 59(18): p. 9870-9876.

[165] Grun, F. and B. Blumberg, Environmental obesogens: organotins and endocrine disruption via nuclear receptor signaling. *Endocrinology*, 2006. 147(6 Suppl): p. S50-5.

[166] Janesick, A. and B. Blumberg, Obesogens, stem cells and the developmental programming of obesity. *Int. J. Androl.*, 2012. 35(3): p. 437-48.

[167] Power, C. and B. J. Jefferis, Fetal environment and subsequent obesity: a study of maternal smoking. *Int. J. Epidemiol.*, 2002. 31(2): p. 413-9.

[168] Desai, M. and C. N. Hales, Role of fetal and infant growth in programming metabolism in later life. *Biol. Rev. Camb. Philos. Soc.*, 1997. 72(2): p. 329-48.

[169] Hales, C. N., M. Desai, and S. E. Ozanne, The Thrifty Phenotype hypothesis: how does it look after 5 years? *Diabet. Med.*, 1997. 14(3): p. 189-95.

[170] McMillen, I. C. and J. S. Robinson, Developmental origins of the metabolic syndrome: prediction, plasticity, and programming. *Physiol. Rev.*, 2005. 85(2): p. 571-633.

[171] Ferezou-Viala, J., et al., Long-term consequences of maternal high-fat feeding on hypothalamic leptin sensitivity and diet-induced obesity in the offspring. *Am. J. Physiol. Regul. Integr. Comp. Physiol.*, 2007. 293(3): p. 1056-1062.

[172] Howie, G. J., et al., Maternal nutritional history predicts obesity in adult offspring independent of postnatal diet. *J. Physiol.*, 2009. 587(Pt 4): p. 905-915.

[173] Samuelsson, A. M., et al., Diet-induced obesity in female mice leads to offspring hyperphagia, adiposity, hypertension, and insulin resistance: a novel murine model of developmental programming. *Hypertension*, 2008. 51(2): p. 383-92.

[174] Pochin, E. E. and F. F. Rundle, Deposition of adipose tissue between ocular muscle fibres in thyrotoxicosis. *Clin. Sci.* (Lond), 1949. 8(1-2): p. 89-95.

[175] Meunier, P., et al., [Osteoporosis and adipose involution of cell population of the marrow. Quantitative study of 51 iliac crest biopsies]. *Presse Med.*, 1970. 78(12): p. 531-4.

[176] Das, U. N., Obesity, metabolic syndrome X, and inflammation. *Nutrition*, 2002. 18(5): p. 430-2.

[177] Gupta, A. and V. Gupta, Metabolic syndrome: what are the risks for humans? *Biosci. Trends*, 2010. 4(5): p. 204-12.

[178] Weisberg, S. P., et al., Obesity is associated with macrophage accumulation in adipose tissue. *J. Clin. Invest.*, 2003. 112(12): p. 1796-1808.

[179] Ortega Martinez de Victoria, E., et al., Macrophage content in subcutaneous adipose tissue: associations with adiposity, age, inflammatory markers, and whole-body insulin action in healthy Pima Indians. *Diabetes*, 2009. 58(2): p. 385-393.

[180] Lumeng, C. N., J. L. Bodzin, and A. R. Saltiel, Obesity induces a phenotypic switch in adipose tissue macrophage polarization. *J. Clin. Invest.*, 2007. 117(1): p. 175-184.

[181] Aron-Wisnewsky, J., et al., Human adipose tissue macrophages: m1 and m2 cell surface markers in subcutaneous and omental depots and after weight loss. *J. Clin. Endocrinol. Metab.*, 2009. 94(11): p. 4619-23.

[182] Shah, A., N. Mehta, and M. P. Reilly, Adipose inflammation, insulin resistance, and cardiovascular disease. *JPEN J. Parenter Enteral Nutr.*, 2008. 32(6): p. 638-44.

[183] McGillicuddy, F. C., et al., Interferon gamma attenuates insulin signaling, lipid storage, and differentiation in human adipocytes via activation of the JAK/STAT pathway. *J. Biol. Chem.*, 2009. 284(46): p. 31936-31944.

[184] Goedecke, J. H., et al., Reduced gluteal expression of adipogenic and lipogenic genes in Black South African women is associated with obesity-related insulin resistance. *J. Clin. Endocrinol. Metab.*, 2011. 96(12): p. 2029-2033.

[185] Buren, J. and J. W. Eriksson, Is insulin resistance caused by defects in insulin's target cells or by a stressed mind? *Diabetes Metab. Res. Rev.*, 2005. 21(6): p. 487-94.

[186] Kotelevtsev, Y., et al., 11beta-hydroxysteroid dehydrogenase type 1 knockout mice show attenuated glucocorticoid-inducible responses and resist hyperglycemia on obesity or stress. *Proc. Natl. Acad. Sci. USA*, 1997. 94(26): p. 14924-14929.

[187] Agwunobi, A. O., et al., Insulin resistance and substrate utilization in human endotoxemia. J Clin Endocrinol Metab, 2000. 85(10): p. 3770-8.

[188] Pei, L. and P. Tontonoz, Fat's loss is bone's gain. *J. Clin. Invest.*, 2004. 113(6): p. 805-806.

[189] Sugimura, R. and L. Li, Shifting in balance between osteogenesis and adipogenesis substantially influences hematopoiesis. *J. Mol. Cell Biol.*, 2010. 2(2): p. 61-2.

[190] Hausman, G. J. and L. R. Richardson, Histochemical and ultrastructural analysis of developing adipocytes in the fetal pig. *Acta Anat.* (Basel), 1982. 114(3): p. 228-47.

[191] Richardson, R. L., et al., Transforming growth factor type beta (TGF-beta) and adipogenesis in pigs. *J. Anim. Sci.*, 1989. 67(8): p. 2171-2180.

[192] Crandall, D. L., et al., Autocrine regulation of human preadipocyte migration by plasminogen activator inhibitor-1. *J. Clin. Endocrinol. Metab.*, 2000. 85(7): p. 2609-14.

[193] Fukumura, D., et al., Paracrine regulation of angiogenesis and adipocyte differentiation during in vivo adipogenesis. *Circ. Res.*, 2003. 93(9): p. e88-97.

[194] Rubina, K., et al., Adipose stromal cells stimulate angiogenesis via promoting progenitor cell differentiation, secretion of angiogenic factors, and enhancing vessel maturation. *Tissue Eng. Part A*, 2009. 15(8): p. 2039-50.

[195] Hausman, G. J. and R. L. Richardson, Adipose tissue angiogenesis. *J. Anim. Sci.*, 2004. 82(3): p. 925-34.

[196] Beck, C. S., The Development of a New Blood Supply to the Heart by Operation. *Ann. Surg.*, 1935. 102(5): p. 801-13.

[197] Suga, H., et al., Adipose tissue remodeling under ischemia: death of adipocytes and activation of stem/progenitor cells. *Plast. Reconstr. Surg.*, 2010. 126(6): p. 1911-23.

[198] Villaret, A., et al., Adipose tissue endothelial cells from obese human subjects: differences among depots in angiogenic, metabolic, and inflammatory gene expression and cellular senescence. *Diabetes*, 2010. 59(11): p. 2755-63.

[199] Beresford, J. N., et al., Evidence for an inverse relationship between the differentiation of adipocytic and osteogenic cells in rat marrow stromal cell cultures. *J. Cell Sci.*, 1992. 102 (Pt 2): p. 341-51.

[200] Takahashi, K., et al., Modification of strain-specific femoral bone density by bone marrow-derived factors administered neonatally: a study on the spontaneously osteoporotic mouse, SAMP6. *Bone Miner.*, 1994. 24(3): p. 245-255.

[201] Kajkenova, O., et al., Increased adipogenesis and myelopoiesis in the bone marrow of SAMP6, a murine model of defective

osteoblastogenesis and low turnover osteopenia. *J. Bone Miner. Res.*, 1997. 12(11): p. 1772-1779.

[202] Cui, Q., G. J. Wang, and G. Balian, Steroid-induced adipogenesis in a pluripotential cell line from bone marrow. *J. Bone Joint Surg. Am.*, 1997. 79(7): p. 1054-63.

[203] Cui, Q., G. J. Wang, and G. Balian, Pluripotential marrow cells produce adipocytes when transplanted into steroid-treated mice. Connect Tissue Res, 2000. 41(1): p. 45-56.

[204] Cui, Q., et al., The Otto Aufranc Award. Lovastatin prevents steroid induced adipogenesis and osteonecrosis. *Clin. Orthop. Relat. Res.*, 1997(344): p. 8-19.

[205] Li, X., et al., Lovastatin inhibits adipogenic and stimulates osteogenic differentiation by suppressing PPARgamma2 and increasing Cbfa1/Runx2 expression in bone marrow mesenchymal cell cultures. *Bone*, 2003. 33(4): p. 652-9.

[206] Wang, Y., et al., Alcohol-induced adipogenesis in bone and marrow: a possible mechanism for osteonecrosis. *Clin. Orthop. Relat. Res.*, 2003(410): p. 213-24.

[207] Akune, T., et al., PPARgamma insufficiency enhances osteogenesis through osteoblast formation from bone marrow progenitors. *J. Clin. Invest.*, 2004. 113(6): p. 846-55.

[208] Moerman, E. J., et al., Aging activates adipogenic and suppresses osteogenic programs in mesenchymal marrow stroma/stem cells: the role of PPAR-gamma2 transcription factor and TGF-beta/BMP signaling pathways. *Aging Cell*, 2004. 3(6): p. 379-389.

[209] Liu, T. M. and E. H. Lee, Transcriptional regulatory cascades in runx2-dependent bone development. *Tissue Eng. Part B Rev.*, 2013. 19(3): p. 254-63.

[210] Gong, Y., et al., LDL receptor-related protein 5 (LRP5) affects bone accrual and eye development. *Cell*, 2001. 107(4): p. 513-23.

[211] Loke, Y. K., C. S. Kwok, and S. Singh, Comparative cardiovascular effects of thiazolidinediones: systematic review and meta-analysis of observational studies. *BMJ*, 2011. 342.

[212] Nissen, S. E. and K. Wolski, Rosiglitazone revisited: an updated meta-analysis of risk for myocardial infarction and cardiovascular mortality. *Arch. Intern. Med.*, 2010. 170(14): p. 1191-1201.

[213] Harmon, G. S., M. T. Lam, and C. K. Glass, PPARs and lipid ligands in inflammation and metabolism. *Chem. Rev.*, 2011. 111(10): p. 6321-40.

[214] Stagakis, I., et al., Anti-tumor necrosis factor therapy improves insulin resistance, beta cell function and insulin signaling in active rheumatoid arthritis patients with high insulin resistance. *Arthritis Res. Ther.*, 2012. 14(3).

[215] Goldfine, A. B., et al., Use of salsalate to target inflammation in the treatment of insulin resistance and type 2 diabetes. *Clin. Transl. Sci.*, 2008. 1(1): p. 36-43.

[216] Beranger, G. E., et al., In vitro brown and "brite"/"beige" adipogenesis: Human cellular models and molecular aspects. *Biochim. Biophys. Acta*, 2013. 1831(5): p. 905-14.

[217] Nedergaard, J., T. Bengtsson, and B. Cannon, Unexpected evidence for active brown adipose tissue in adult humans. *Am. J. Physiol. Endocrinol. Metab.*, 2007. 293(2): p. 444-452.

[218] Ouellet, V., et al., Brown adipose tissue oxidative metabolism contributes to energy expenditure during acute cold exposure in humans. *J. Clin. Invest.*, 2012. 122(2): p. 545-552.

[219] Lu, X., et al., Resistance to obesity by repression of VEGF gene expression through induction of brown-like adipocyte differentiation. *Endocrinology*, 2012. 153(7): p. 3123-32.

[220] Zilberfarb, V., et al., Human immortalized brown adipocytes express functional beta3-adrenoceptor coupled to lipolysis. *J. Cell Sci.*, 1997. 110 (Pt 7): p. 801-7.

Index

#

21st century, ix, 95, 96, 165

A

access, 61, 69, 152, 164
acetylation, 167
acid, 14, 23, 100, 112, 121, 125, 131, 151, 158, 163, 175
acidic, 150
active site, 133
AD, 39, 40, 41, 53, 55, 64, 107
adaptation(s), 128, 136
adenine, 134, 143
adhesion, 150
adiponectin, 14, 62, 64, 112, 158
adiposity, 80, 109, 112, 113, 115, 117, 119, 122, 130, 139, 151, 152, 160, 167, 177, 178
ADP, 166
adult stem cells, 89
adulthood, 128
adults, 66, 67, 82, 163
adverse effects, 6, 48
aesthetic, 48
age, ix, x, 16, 48, 55, 63, 67, 76, 95, 98, 103, 110, 111, 143, 149, 160, 161, 178
aging process, 47, 48, 76
agonist, 115, 131, 154, 171
AIDS, 38, 79, 83, 85, 87, 91
allele, 58, 59, 60, 61
alopecia, 46
alters, ix, 29, 82, 109
AME, 104
amino acid(s), 42, 47, 50, 154
angiogenesis, 20, 37, 161, 179
anisotropy, 167
antagonism, 114, 171
antibody, 117, 118, 120, 122, 123, 125
antioxidant, 129, 130, 133, 138, 139, 158
antiretrovirals, 48
antisense, 55, 87
APC, 9, 10
APL, 41
aplastic anemia, 156, 173
apoptosis, 87, 107, 114, 121, 125, 143, 156, 158, 171, 173, 176
appetite, 152, 159
arginine, 63, 104
arrest, 11, 140
arsenic, xi, 146
arteries, 56
arteriosclerosis, 41
arthritis, 155
aspartate, 173
atherosclerosis, viii, 36, 38, 39, 44, 47, 154
ATP, 65, 137, 148, 166
atrophy, 46

attachment, 13
autonomic nervous system, 37
autosomal dominant, 26, 29, 44, 55, 57, 64, 78
autosomal recessive, 45, 57, 86, 88

B

BAC, 55
bacterial artificial chromosome, 55, 56
barriers, 149
basal metabolic rate, 40
basic research, 75
BD, 140
beverages, 139
Bible, 165
biochemistry, 121
biological activity(s), 103, 105
biological processes, 37
biomaterials, 6
biomechanics, 167
biopsy, 69
biotechnology, 122
birth weight, 159
black women, 160
blood, 5, 21, 37, 97, 147, 161
blood flow, 37
blood vessels, 161
BMI, 101, 152, 154, 160
body fat, 36, 38, 82, 89, 169
body weight, 47, 59, 61, 130, 135, 161, 167
bone form, ix, 2, 6, 11, 12, 15, 16, 17, 20, 22, 26, 27, 28, 30, 32, 33, 96, 110, 114, 116, 117, 118, 119, 120, 121, 122, 123, 124
bone growth, 23, 122, 123, 125, 158
bone marrow, ix, 2, 5, 20, 21, 22, 23, 24, 26, 29, 32, 33, 34, 66, 73, 101, 102, 109, 110, 111, 114, 115, 117, 119, 120, 122, 123, 124, 125, 148, 150, 161, 179, 180
bone mass, x, 7, 9, 12, 17, 18, 21, 25, 28, 32, 109, 117, 121, 122, 123, 125, 162, 172
bone resorption, 11, 27
bone volume, 115
brain, 29, 48, 138, 152
brain functions, 29
breakdown, 134
breast cancer, 25, 115, 122
buffalo, 48
building blocks, 165

C

Ca^{2+}, 9
calcification, 17, 18, 41
caloric intake, 146, 147
caloric restriction, 139
calorie, 152, 159
cancer, ix, xi, 8, 24, 25, 29, 76, 97, 109, 110, 111, 114, 115, 119, 120, 122, 124, 146, 164
cancer cells, 25
cancer therapy, 114
capillary, 157, 161
carbohydrate(s), 134, 154
carbohydrate metabolism, 154
carbon, 42
carboxyl, 42, 54, 60, 80, 85
carcinoma, 25
cardiomyopathy, 39, 40, 44, 52, 53, 58, 86, 144
cardiovascular disease(s), 39, 44, 147, 151, 154, 178
cartilage, 2, 16, 31, 39, 66, 148, 164
cascades, 180
catabolism, 144
catecholamines, 161
cattle, 146
Caucasian population, 170
Caucasians, 170
cell biology, 124
cell culture, 20, 23, 99, 103, 116, 133, 149, 158, 179, 180
cell cycle, 42, 130, 167, 168
cell differentiation, vii, 8, 12, 15, 42, 62, 68, 83, 140, 179
cell fate, 8, 28, 29, 30, 149, 150, 164
cell line(s), x, 5, 6, 8, 14, 15, 16, 21, 25, 32, 47, 67, 69, 83, 91, 99, 100, 102, 105,

110, 111, 114, 118, 122, 148, 149, 150, 152, 163, 164, 166, 180
cell signaling, 7, 141
cell size, 160
cell surface, 9, 16, 155, 178
cellular biology, x, 145, 164
cellular differentiation, vii
cellular energy, 131
cellular homeostasis, x, 127, 133
cerebrovascular disease, 47
CGL, 39, 57
challenges, 163
chemical(s), ix, 62, 65, 96, 99, 100, 101, 102, 103, 105, 158
chemokines, 160
chemotaxis, 174
chemotherapeutic agent, 115
chemotherapy, ix, 109, 110, 111, 112, 114, 115, 119, 121, 122, 125
childhood, 40
children, 22, 47, 82, 87, 121, 122
chimera, 23
cholesterol, viii, 35, 38, 96
chondrocyte, viii, 1, 8, 24, 27, 99
chromatography, 106, 107
chromosome, 47, 78
chronic illness, 38
chronic renal failure, 39
cilia, 13, 29
cilium, 13
cirrhosis, 48
classes, 41
classification, 38, 57
cleavage(s), 50, 58
cleidocranial dysplasia, ix, 25, 96
clinical application, 11
clinical problems, xi, 146
clinical syndrome, viii, 35
clinical trials, 6, 49, 75
cloning, 24, 82, 165
coding, 82
codon, 60
cognition, 97
cognitive function, ix, 95, 97, 106
colitis, 24
collaboration, 156
collagen, 22, 56, 111, 150
colon, 25
colon cancer, 25
communication, 167
complement, 37, 41
complex interactions, ix, 109
complications, viii, 6, 35, 38, 44, 45, 46, 47, 48, 62, 63, 90
composition, 85, 87, 111, 114, 119, 149, 172
compounds, 60, 65, 102, 130, 170
Concise, 19, 22
conduction, ix, 36, 81
conductive hearing loss, 47
configuration, 132
connective tissue, 37, 147, 148, 161
consensus, 168
conservation, 29
constituents, 137
control group, 98
controversial, 105
cooperation, 25, 155
copolymer, 104
correlation, 5
cortisol, 50, 161
CT, 31, 66, 163
CT scan, 31, 163
culture, 2, 3, 6, 15, 69, 70, 74, 99, 116, 130, 148, 150, 151, 162, 166, 167
culture conditions, 15, 70
culture medium, 150
cure, 38
cycles, 102, 156
cycling, 115
cyclophosphamide, 122
cysteine, 42, 59, 130, 132, 150
cytochrome, 134, 135, 137, 143, 144
cytokine(s), vii, viii, 2, 3, 7, 16, 18, 37, 49, 50, 83, 147, 157, 160, 170
cytoplasm, 57, 113
cytoskeleton, 56, 149, 167

D

damages, 125
decomposition, 143
defects, ix, 6, 11, 18, 23, 25, 36, 38, 44, 47, 53, 56, 57, 62, 81, 84, 86, 89, 91, 92, 96, 100, 114, 115, 117, 122, 162, 178
defence, 130
deficiency, ix, 11, 36, 56, 57, 70, 78, 81, 84, 117, 120, 121, 175
degradation, 113, 151
Delta, 171
dementia, 97
dental implants, 21
deposition, 53
deposits, 48, 62, 100, 147, 159, 163
depression, 41
depth, vii
deregulation, 36, 113, 115, 116, 117, 119, 120
derivatives, 103
dermis, 148
detectable, 80
detection, 47
detoxification, 129, 158
developing countries, 96
developmental process, 66
diabetes, xi, 41, 57, 80, 81, 82, 97, 106, 141, 146, 152, 154, 157, 160, 163, 168
diabetic patients, 171
dichotomy, vii, viii, 2, 19
diet, 24, 130, 135, 140, 144, 152, 158, 160, 163, 174, 176, 177
dietary fiber, 158
dilated cardiomyopathy, 81, 86
dimerization, 62, 129, 138
disability, 97
disease model, ix, 36, 70, 75, 154
diseases, viii, 9, 35, 36, 43, 50, 51, 52, 60, 64, 74, 75, 76, 79, 97, 123, 147, 164, 171
disorder, 38, 44, 45, 52, 57
displacement, 155
distribution, 31, 40, 44, 64, 71, 78, 169
divergence, 29
diversity, 73, 86
DNA, 8, 16, 24, 30, 31, 32, 42, 45, 47, 49, 74, 129, 138, 142, 150, 154, 157, 168, 169
DNA damage, 42, 129, 142
DNA repair, 47
dosage, 18
down-regulation, 45, 139, 157, 168, 176
Drosophila, 8, 12, 14, 24, 29, 30
drug discovery, 73
drugs, xi, 38, 49, 87, 100, 136, 146, 164, 171
dyslipidemia, 39, 40, 41, 46, 159
dysplasia, 45, 46, 57, 77, 79, 81, 87

E

early postnatal development, 55, 62
ECM, 53, 55, 58
education, 97
elderly population, 97, 106
electric current, 19
electric field, 19
electron(s), 132, 134, 135, 137, 157, 163
elucidation, vii, viii, 2, 19, 76
embryogenesis, 12
embryonic stem cells, 42, 52, 100, 149
encephalomyelitis, 155
encoding, 72, 78, 85, 86, 89
endocrine, 36, 37, 65, 77, 82, 83, 84, 147, 165, 177
endocrine system, 83
endocrinology, 120, 142, 159
endothelial cells, 3, 44, 161, 179
endothelial dysfunction, 143
endotoxemia, 178
energy, 16, 31, 37, 65, 66, 82, 88, 89, 128, 131, 134, 135, 137, 139, 147, 148, 159, 163, 165, 166, 172, 181
energy expenditure, 17, 66, 163, 181
engineering, 6, 16
environment, ix, x, 97, 109, 115, 128, 159, 160, 177
environmental factors, vii, 152
enzymatic activity, 135

enzyme(s), 37, 49, 59, 132, 133, 135, 137, 138, 143, 155, 161
epidemic, 38, 147, 164
epithelial cells, 147
erythroid cells, 24
ester, 42, 104, 129
estrogen, 37, 84, 98, 102, 117, 120
etiology, 38, 49
Europe, 159
evidence, x, 5, 8, 12, 25, 42, 49, 75, 86, 97, 127, 137, 156, 161, 174, 181
evil, 141
examinations, 97
exposure, 16, 17, 47, 67, 115, 116, 167, 181
external influences, 37
extracellular matrix, 55, 56, 58, 72, 83, 85, 111, 149, 151, 168
extracts, 104, 105

F

FAD, 134, 137, 143
families, x, 9, 89, 146
famine, 164
fatty acids, 14, 131, 160
FDA, 16
fetus, 159
fibers, 159
fibrin, 21, 33
fibroblasts, 44, 52, 53, 56, 57, 58, 62, 69, 70, 71, 72, 73, 74, 75, 79, 80, 90, 92, 100, 148, 171, 172, 173, 175
fibrosis, 174
filament, 41, 56
fistulas, 6
food, 65, 146, 148, 156, 164
food intake, 65, 148
force, 53, 56
formation, x, 7, 9, 14, 15, 16, 17, 18, 24, 30, 51, 68, 100, 109, 112, 117, 118, 119, 125, 130, 140, 141, 151, 157, 165, 167, 180
fractures, 54, 59, 78
free radicals, 158
FRP, 117
fruits, 146
functional analysis, 26
fusion, 18

G

gastric mucosa, 24
gastrointestinal tract, 171
gastrulation, 148
gender differences, 102
gene expression, 3, 4, 9, 11, 13, 17, 21, 25, 34, 42, 52, 64, 68, 71, 84, 91, 105, 116, 123, 125, 155, 167, 168, 169, 171, 174, 176, 179, 181
gene promoter, 62, 129
gene targeting, 52, 58, 60, 90
gene therapy, 33
gene transfer, 32, 74, 87
genetic defect, 38
genetic mutations, 45
genetics, 75, 102, 122, 160
genome, 14, 30, 68, 155, 157, 169, 171
genomic instability, 47
genotype, 84
gestational diabetes, 44
ginger, 158
ginseng, xi, 146
glioblastoma, 13
glioma, 10
glomerulonephritis, 41
glucocorticoid(s), ix, 23, 37, 109, 111, 112, 116, 119, 121, 122, 123, 125, 149, 157, 162, 172, 178
glucocorticoid receptor, 23, 37, 162
glucose, viii, 31, 35, 38, 57, 62, 65, 88, 131, 140, 156, 158, 171, 173, 176
GLUT, 157
GLUT4, 173
glutathione, 138, 157
glycerol, 112
glycogen, 28, 129
glycoproteins, 155
God, 146
grades, 60
granulopoiesis, 120

graph, 98
growth, vii, 2, 4, 5, 8, 18, 19, 22, 24, 25, 34, 39, 40, 45, 52, 53, 58, 62, 105, 121, 129, 137, 141, 149, 150, 151, 161, 166, 177, 179
growth arrest, 129, 149, 150, 151
growth factor, vii, 4, 8, 18, 34, 121, 141, 179
guardian, 138

H

HAART, 49, 50, 63
hair, 39, 47
hair loss, 39
half-life, 128
harmony, 147
healing, 6, 33, 124, 150, 161, 164
health, 38, 76, 96, 106, 147, 160
health care, 96
hematology, 121
hematopoietic stem cells, 24
hemoglobin, 97
hepatomegaly, 39, 40
heterochromatin, 44, 52, 71
heterogeneity, 3, 38
high density lipoprotein, 97
high fat, 159, 160, 175
highly active antiretroviral therapy, 49
histone, x, 28, 146, 155, 157
histone deacetylase, 155
history, 122, 164, 177
human immunodeficiency virus (HIV), 38, 40, 44, 48, 49, 50, 63, 64, 70, 77, 79, 80, 81, 82, 83, 84, 86, 87, 91
HIV/AIDS, 38
HIV-1, 49, 50, 77, 91
homeostasis, viii, 9, 23, 31, 35, 45, 56, 65, 70, 72, 88, 137, 147, 166, 172
hormonal control, 63, 167
hormone(s), 7, 23, 37, 68, 98, 99, 102, 132, 155, 161, 165
human body, vii, 74
human condition, 164
human health, 164
human skin, 85, 175
human subjects, 139, 179
hydrogen, 129, 141, 142, 176
hydrogen peroxide, 129, 141, 142, 176
hydroxyapatite, 6, 22
hyperandrogenism, 44
hyperglycaemia, 41
hyperglycemia, ix, 36, 57, 158, 161, 178
hyperinsulinemia, 40
hyperlipidemia, 40, 97, 151
hypermethylation, 25
hyperplasia, 24, 137, 160
hypertension, viii, xi, 36, 38, 40, 41, 146, 159, 170, 177
hypertriglyceridemia, viii, 35, 38, 48, 158
hypertrophy, 24, 137, 160, 162
hypoglycemia, 155
hypoplasia, 39, 40, 41, 45, 58
hypothermia, 65
hypothesis, 56, 59, 60, 61, 65, 67, 70, 76, 130, 177
hypoxia, 136, 137, 142, 144, 157, 160, 161, 174
hypoxia-inducible factor, 136

I

ideal, 73, 151
identification, 56, 75, 106, 107, 170
identity, 74
iliac crest, 177
immune function, 37, 147
immunomodulatory, 5
improvements, 154
in vitro, x, 4, 5, 6, 9, 12, 14, 15, 17, 21, 23, 29, 32, 33, 34, 49, 50, 63, 64, 66, 71, 75, 84, 100, 102, 114, 115, 116, 122, 125, 130, 135, 145, 150, 151, 162, 166, 169, 170, 175, 177
in vivo, x, 5, 6, 9, 12, 14, 15, 17, 20, 23, 28, 29, 32, 33, 49, 50, 59, 61, 62, 64, 66, 69, 73, 75, 87, 100, 118, 122, 130, 133, 135, 138, 142, 145, 149, 150, 151, 154, 162, 163, 169, 179
incidence, 59

Indians, 178
individuals, 42, 44, 47, 66, 73, 74, 164
inducer, 14
induction, 11, 15, 16, 24, 32, 33, 69, 71, 80, 125, 130, 137, 149, 150, 155, 156, 158, 167, 175, 176, 181
infants, 66
infection, 6, 38, 48, 49, 82, 91
infertility, 44
inflammation, 33, 83, 125, 147, 150, 158, 160, 163, 174, 176, 177, 178, 180, 181
inflammatory bowel disease, 155
inflammatory disease, 163
inflammatory markers, xi, 146, 178
infliximab, 163
inhibition, 9, 11, 13, 14, 31, 49, 50, 58, 71, 73, 76, 85, 105, 113, 118, 120, 123, 124, 131, 133, 137, 140, 148, 149, 156, 161, 168
inhibitor, 11, 14, 16, 28, 50, 55, 59, 70, 73, 75, 78, 79, 81, 82, 92, 113, 114, 118, 121, 129, 133, 161, 168, 179
initiation, x, 127, 134, 149
injury(s), ix, 80, 97, 109, 119, 155, 175
innate immunity, 161
insertion, 8
insulin resistance, viii, 24, 36, 38, 39, 40, 46, 48, 50, 53, 55, 57, 63, 77, 79, 141, 151, 157, 159, 160, 161, 164, 169, 170, 174, 176, 177, 178, 181
insulin sensitivity, 14, 17, 56, 68, 130, 133, 139, 141, 151, 157, 158, 160, 161
insulin signaling, 139, 140, 141, 142, 143, 160, 178, 181
integration, viii, 2, 4, 5, 8, 28, 82
integrity, 52, 56, 58, 73, 88, 90, 91, 138
interference, 18, 131
interferon, 156
intervention, 6
intron, 59
involution, 177
ions, 144
irradiation, ix, 109, 112, 115, 124, 125
ischemia, 179
isolation, 163, 164
Italy, 127

J

Japan, 95, 97

K

kaempferol, 34
ketoacidosis, 40
kidney(s), 48, 164
kinase activity, 133

L

L-arginine, 104, 106
later life, 177
LDL, 26, 27, 180
lead, 36, 44, 48, 80, 149, 152, 155, 156
legs, 41, 48
leptin, 14, 37, 62, 64, 83, 112, 147, 151, 152, 177
leucine, 58, 155
leukemia, 7, 24
leukocytes, 24, 121, 160
life expectancy, 97
life sciences, 124
ligament, 2
ligand, ix, 5, 11, 13, 16, 96, 100, 105, 113, 154, 155, 171
light, 173
lipid metabolism, 49, 65, 71, 161, 164
lipids, 60, 61, 65, 147, 148
lipodystrophy, ix, 36, 38, 39, 40, 41, 44, 45, 46, 47, 48, 49, 50, 51, 54, 55, 57, 58, 60, 62, 63, 64, 65, 70, 71, 72, 73, 75, 76, 77, 78, 79, 80, 83, 84, 85, 86, 87, 89, 90, 91, 96, 154, 170
lipolysis, 20, 63, 84, 92, 106, 156, 168, 169, 173, 174, 181
lithium, 11, 28
liver, 48, 57, 106, 155, 171
liver failure, 155
localization, 61, 63, 79

loci, 152
locus, 78, 86, 101
longevity, 59, 90
longitudinal study, 125
low-density lipoprotein, 9
luciferase, 14
lumbar spine, 118
lung cancer, 25
Luo, 33, 112, 123
lymphocytes, 85
lymphoid, 8, 9, 10, 24
lysine, 28

M

machinery, vii, x, 127, 128, 134, 137
macrophages, 49, 143, 154, 160, 178
magnetic field, 19
majority, 5, 149, 164
mammalian cells, 132
mammalian tissues, 56
mammals, 7, 52, 65, 137
man, 175
manipulation, 61, 87
marrow, ix, 5, 20, 23, 33, 109, 110, 111, 112, 113, 114, 115, 117, 119, 120, 122, 124, 125, 166, 177, 179, 180
mass, 10, 26, 54, 117, 128, 133, 135, 152, 162
maternal smoking, 159, 177
matrix, 4, 32, 37, 72, 111, 150, 163, 167, 168, 171, 175
matrix metalloproteinase, 168, 171, 175
matter, 65
McGillicuddy, 178
mechanical stress, 46, 53, 56
medical, 96, 112
medicine, 125
mellitus, 41, 46
menopause, 102, 112, 117
mental retardation, 39
mesenchymal stem cells, ix, x, 3, 5, 19, 20, 21, 22, 23, 26, 27, 28, 29, 32, 33, 34, 36, 37, 47, 65, 68, 69, 82, 86, 88, 95, 98, 102, 110, 121, 123, 124, 132, 138, 140, 149, 163, 173, 176, 177
mesenchyme, 172
mesoderm, 148
messenger ribonucleic acid, 173
messengers, 132
meta-analysis, 139, 142, 171, 180
Metabolic, 46, 159, 166, 174, 178
metabolic disorder(s), 36, 48, 65
metabolic pathways, 172
metabolic syndrome, xi, 78, 85, 86, 97, 146, 151, 157, 159, 160, 163, 169, 174, 175, 177
metabolism, 4, 26, 37, 66, 80, 82, 83, 89, 97, 106, 121, 125, 131, 134, 139, 142, 147, 156, 165, 177, 180, 181
metabolites, 100, 134
metabolized, 137
metalloproteinase, 46, 58, 77, 78, 80, 87
methyl cellulose, 158
methylation, 16, 31, 157
methylcellulose, 149
microbiota, 152
migration, 6, 42, 161, 179
mineralization, 4, 11, 18, 23, 75, 116
mitochondria, x, 66, 103, 127, 133, 134, 135, 138, 144, 148, 166
mitochondrial toxins, 137
mitogen, 129, 157, 168
mitosis, 47, 79, 148, 150
models, vii, viii, x, 4, 14, 36, 52, 53, 58, 59, 62, 63, 64, 69, 70, 73, 75, 76, 89, 115, 117, 118, 122, 145, 148, 151, 152, 154, 155, 164, 166, 181
modifications, 13, 42, 63
molecular medicine, 165
molecular oxygen, 134, 143
molecules, 5, 113, 123, 130, 133, 150
monoclonal antibody, 27, 117, 118, 123, 124
morbidity, 2
morphology, 5, 50, 52, 123, 138
mortality, 180
Moses, 81
motif, 18, 42, 60

MR, 125, 138, 142
mRNA(s), 42, 50, 55, 57, 61, 82, 87, 106, 115, 116, 151
multiple myeloma, 118, 125
multiple sclerosis, 88, 155
multipotent, vii, 1, 2, 3, 6, 14, 20, 22, 66, 81, 88, 93, 99, 125
muscles, 48
muscular dystrophy, ix, 36, 40, 44, 52, 53, 54, 55, 57, 58, 78, 83, 90
muscular tissue, 56
musculoskeletal, 121
mutagenesis, 90, 106
mutant, 28, 50, 52, 56, 58, 59, 60, 61, 63, 64, 82, 85, 154
mutation(s), viii, ix, 10, 26, 27, 29, 35, 38, 41, 43, 44, 45, 46, 47, 50, 52, 55, 56, 57, 58, 61, 62, 63, 64, 76, 77, 78, 79, 81, 84, 86, 88, 89, 91, 92, 96, 101, 154, 156, 162, 167, 169, 170
myoblasts, 156
myocardial infarction, 47, 163, 180
myocardium, 161
myocyte, 67, 99

N

NAD, 131, 134, 135, 140
NADH, 134, 135
nanoparticles, 33
natural compound, 130
necrosis, 85
neonates, 65
nerve, 37
Netherlands, 91
neurodegeneration, 9, 26
neurodegenerative diseases, 76
neurogenesis, 8
neurons, 24
neutrophils, 143
New England, 125
New Zealand, 152
NH_2, 9
nicotinamide, 134, 143
nitric oxide, 104, 138
nitric oxide synthase, 104
Nobel Prize, 74
nodules, 46
normal aging, 48
normal development, 62
Nrf2, 157, 175
NRT, 49, 64
nuclear membrane, viii, 35, 41, 43, 47, 52, 82
nuclear receptors, 154
nuclear surface, 56
nuclei, 54, 59, 71, 79, 80
nucleoside reverse transcriptase inhibitors, 49, 64
nucleotides, 60
nucleus, 9, 10, 25, 42, 57, 81, 113, 148
null, 8, 52, 53, 56, 57, 58, 71, 100, 150, 154, 167
nutrient(s), 128, 133, 134, 135, 136, 137, 138, 142, 159
nutrition, 96

O

obstacles, 69
oil, 125, 148, 151
old age, 58, 76
oncoproteins, 60
organ(s), 2, 3, 36, 65, 77, 80, 84, 117, 147, 148, 165, 166
ossification, 30
osteoblastgenesis, ix, 95, 98, 99, 100, 101, 102, 103
osteoclastogenesis, 125
osteocyte, 114, 121
Osteogenesis, 2, 111, 123, 161
osteogenesis imperfecta, 6, 22
osteogenic, vii, viii, x, 1, 5, 6, 7, 8, 9, 10, 12, 14, 15, 16, 17, 18, 20, 21, 22, 23, 26, 29, 30, 32, 33, 34, 101, 110, 111, 112, 113, 114, 115, 116, 118, 124, 162, 164, 179, 180
osteoporosis, xi, 9, 27, 32, 58, 97, 102, 113, 116, 117, 118, 120, 122, 123, 125, 146, 159, 162, 176

ovariectomy, 117
overnutrition, 159, 160
overweight, 160
ox, 23
oxidation, 132, 133, 135, 141, 143
oxidative damage, x, 128
oxidative stress, 45, 49, 50, 79, 91, 129, 135, 157, 173, 174, 175, 176
oxygen, 134, 135, 138, 141, 142, 144, 157

P

pain, 6
parallel, 64
parathyroid, 19, 118
parathyroid hormone, 19, 118
pathogenesis, 49, 56, 63, 74, 159
pathology, 53, 56, 80, 123
pathophysiological, 43, 50, 75
pathophysiology, viii, 36, 49, 58, 65, 75, 159
pathways, vii, viii, x, 2, 7, 8, 9, 11, 14, 15, 17, 29, 49, 69, 103, 114, 122, 131, 133, 137, 145, 156, 157, 161, 164, 174
PCP, 9, 10
penetrance, 55
peptidase, 42
peptide(s), 37, 141
perinatal, 155
permeability, 135, 176
permission, 153
peroxide, 142
PET, 31, 66, 163
phalanges, 45
pharmacology, 122
phenotype(s), x, 3, 4, 8, 10, 19, 23, 38, 45, 46, 49, 51, 52, 53, 54, 56, 57, 58, 59, 60, 61, 62, 63, 64, 65, 69, 70, 75, 81, 92, 101, 110, 111, 146, 150, 151, 160, 166
phosphate, 6, 22, 112, 143
phosphorylation, 9, 16, 17, 26, 28, 65, 129, 138, 140, 144, 157, 168, 174
photophobia, 47
Physiological, 83
physiological mechanisms, 69
physiology, 20, 82, 121, 122, 125, 169
physiopathology, 52, 57, 60
PI3K, 133, 136
pigmentation, 40, 41
pigs, 179
placebo, 27, 124
plants, 130
plaque, 171
plasma levels, 131
plasma membrane, 25, 132
plasmid DNA, 33
plasminogen, 161, 179
plasticity, 177
platform, 2
playing, viii, 36, 65
pleotropic, viii, 2
point mutation, 43, 47, 50, 52, 58, 64, 81
polarity, 9, 10
polarization, 178
pollution, 142
polycystic ovarian syndrome, 41
polymer, 32
polymerase, 172
polypeptide, 13, 90
polyphenols, 158, 175
polyunsaturated fatty acids, 7, 155, 171
pools, 125
population, 2, 19, 87, 96, 97, 177
Portugal, 73, 85
positive feedback, 15
potato, 158, 175
precursor cells, 22, 149, 168
premature death, 52, 54
preparation, 103
preservation, 119
primate, 117, 122
pro-atherogenic, 44
progenitor cells, x, 21, 23, 109, 110, 119, 124, 125, 128, 138, 148, 179
progesterone, 37
programming, 177
pro-inflammatory, 160
prolactin, 83
proliferation, 4, 6, 8, 21, 26, 58, 71, 74, 110, 117, 129, 132, 141, 158, 173

proline, 58, 137, 144
promoter, 57, 64, 113, 154
prosperity, 147
prostaglandins, 103, 163
protease inhibitors, 49, 50, 63, 79, 80, 81, 83, 84, 91
proteasome, 158
protein family, 149, 172
protein folding, xi, 146
protein kinase C, 9
protein kinases, 129
proteins, 5, 7, 8, 9, 13, 21, 30, 31, 32, 37, 38, 41, 42, 45, 47, 56, 68, 72, 76, 82, 83, 100, 106, 112, 132, 147, 148, 149, 150, 151, 156, 162, 166, 173
proteolysis, 47
protoplasm, 147
psoriasis, 155, 171
PTEN, 89, 132, 136, 141, 143
puberty, 39, 40, 44, 57, 63
purification, 104, 105

Q

quality of life, 76

R

race, 160
radiation, 110
radical formation, 157
radicals, 142, 143, 157
radiotherapy, 110, 122
radius, 118
reactive oxygen, x, 127, 138, 140, 143, 144, 174
reagents, 61
receptors, 7, 9, 16, 17, 18, 28, 31, 33, 37, 64, 84, 85, 96, 100, 113, 119, 151, 155, 169, 171
recognition, 47
recombination, 52
reconstruction, 6
recovery, x, 110, 120, 122, 124, 125
red wine, 130
redistribution, viii, 35, 36, 76, 79
regeneration, 4, 22, 31, 106
regenerative medicine, 6, 20
regulatory changes, 4
relevance, xi, 75, 138, 146, 151
remediation, 122
remodelling, vii, 112, 123, 128, 136, 167
renin, 170
repair, 20, 22, 123
replication, 49, 150, 168
repression, 88, 156, 173, 181
repressor, 8, 13, 155
requirements, 134
researchers, vii, 163, 164
reserves, 147
residues, 42, 132
resistance, ix, 36, 39, 40, 46, 57, 89, 131, 141, 147, 152, 157, 160, 163, 178, 181
respiration, 65, 136, 148
response, 17, 19, 30, 32, 42, 79, 83, 84, 133, 136, 139, 148, 161, 173, 174, 175
responsiveness, 17, 33, 64, 132, 174
resveratrol, xi, 130, 139, 146, 158, 176
retardation, 39, 40, 45, 53, 54, 55, 58, 62
reticulum, xi, 80, 146
retinoblastoma, 45
retinol, 112
rheumatoid arthritis, 163, 181
rights, 107
risk(s), viii, x, 6, 36, 38, 44, 48, 76, 82, 84, 109, 110, 116, 147, 151, 159, 160, 163, 164, 178, 180
risk factors, 44
RNA, 61, 131, 172
RNA splicing, 61
RNAi, 14
rodents, 100
root, 24
ROS production, x, 127, 129, 132, 133, 134, 136, 137, 142
rosiglitazone, 171

S

scleroderma, 47
secrete, 65
secretion, 6, 66, 112, 160, 179
sedentary lifestyle, 152
senescence, 45, 48, 50, 71, 79, 88, 179
sensing, 142, 144
sensitivity, 157, 158, 160, 177
serine, 16, 59, 133, 157
serum, 41, 57, 113, 116, 117, 120, 149, 163, 167
sex, 147, 160
sex steroid, 147
sexual dimorphism, 63
shape, 75, 90, 149, 167
sheep, 146
showing, xi, 64, 72, 146, 150, 154, 156, 157
side effects, 101, 118
signal transduction, 12, 13, 140, 157
signaling pathway, vii, viii, x, 1, 8, 11, 13, 15, 16, 21, 23, 29, 65, 70, 98, 110, 112, 113, 115, 116, 117, 119, 124, 173, 180
signalling, 25, 26, 27, 28, 31, 53, 63, 69, 122, 123, 125, 128, 130, 132, 135, 137
signals, 4, 19, 31, 56, 65, 156, 173
siRNA, 9, 131
skeletal muscle, 53, 58, 66, 89
skeleton, 12, 26, 55, 117, 120
skin, 39, 41, 44, 45, 46, 73, 151
small intestine, 32
smoking, 159
smooth muscle, 3, 53, 56, 88, 91, 92
smooth muscle cells, 3, 53, 56
society, 96
sodium, 171
somatic cell, 41, 42, 87
South Africa, 178
Spain, 35
species, x, 3, 63, 64, 127, 138, 140, 141, 143, 144, 157, 164, 174
specter, 146
speculation, 165
spinal cord, 155, 171
spinal cord injury, 171
spinal fusion, 14, 16, 29
spine, 118, 164
Spring, 20
squamous cell, 124
stability, 42, 79
stabilization, 12
starvation, 137, 146, 164
state(s), ix, 22, 37, 62, 65, 67, 68, 84, 109, 110, 113, 114, 119, 128, 131, 135, 155, 157, 160, 173
stem cell differentiation, 8, 80
stem cell lines, 63, 92
stem cells, vii, ix, 1, 2, 4, 5, 6, 12, 19, 20, 21, 22, 23, 33, 34, 36, 65, 66, 67, 70, 73, 75, 76, 77, 81, 86, 87, 90, 93, 99, 102, 110, 114, 124, 125, 131, 138, 148, 162, 165, 177, 180
steroids, 161
stimulation, 80, 133, 142, 163
stimulus, 162
storage, 3, 16, 37, 65, 146, 147, 151, 159, 178
stress, 19, 56, 74, 119, 131, 137, 142, 157, 160, 161, 167, 174, 178
stress response, 19
stroke, 47
stroma, 5, 124, 166, 180
stromal cells, 2, 5, 20, 21, 22, 23, 29, 33, 34, 99, 113, 125, 148, 150, 160, 179
stromal commitment, x, 110, 111, 112
structure, 86, 91, 112, 133, 154
style, 97
styrene, 104
subcutaneous injection, 38
subcutaneous tissue, 62
substitutes, 22
substitution, 58
substrate(s), 37, 54, 59, 104, 132, 137, 178
Sudan, 150
sulfonamides, 124
Sun, 138, 146
supplementation, 125, 129, 176
suppression, 11, 13, 33, 83, 121, 125, 162, 174, 176

survival, 23, 24, 49, 54, 83, 92, 106, 137, 164, 167
susceptibility, 152, 160
symptomatic treatment, 38
symptoms, 61
syndrome, 10, 25, 39, 41, 43, 49, 50, 56, 57, 63, 64, 78, 81, 82, 84, 85, 86, 88, 91, 92, 158, 159, 162, 178
synthesis, 3, 4, 42, 53, 58, 59, 60, 80, 129

T

T cell, 9, 10
T lymphocytes, 171
target, vii, x, 3, 13, 16, 25, 61, 73, 85, 100, 103, 105, 110, 111, 113, 117, 118, 119, 131, 132, 135, 154, 155, 163, 178, 181
T-cell receptor, 24
techniques, 21, 66
technology, 22, 74, 103
teeth, ix, 95, 97, 106
telomere, 79
TEM, 142
temperature, 65, 163
tendons, 55
testis, 12, 28
testosterone, 84
tetrachlorodibenzo-p-dioxin, 175
TGF, 8, 15, 18, 33, 124, 161, 179, 180
therapeutic targets, 27, 117, 123
therapeutic use, 6
therapeutics, 75, 122, 163
therapy, 20, 22, 38, 49, 55, 60, 63, 64, 79, 83, 87, 88, 111, 112, 116, 117, 118, 119, 122, 181
thiazolidinediones, 100, 154, 155, 163, 180
threonine, 16
thyroid, 7
thyrotoxicosis, 177
TIMP, 78
tissue engineering, vii, viii, 2, 19, 21, 22, 33
tissue homeostasis, 8
titanium, 28
TNF, xi, 49, 85, 106, 112, 146, 160, 163, 171
TNF-alpha, 160, 163, 171
TNF-α, 112
tooth, 47
total cholesterol, 97, 131
toxic effect, 59
toxicity, 49, 54, 60, 122
trafficking, 29
transcription factors, vii, viii, 1, 4, 5, 7, 9, 11, 13, 15, 18, 30, 31, 42, 45, 64, 68, 71, 72, 73, 74, 111, 112, 116, 129, 139, 149, 156, 157, 172, 176
transcriptomics, 125
transcripts, 62, 64
transduction, 9, 16, 31, 73
transfection, 33
transformation, 6, 136
transforming growth factor, 15, 25, 161, 166, 175
translation, 19, 42
transmission, 53, 56, 81
transplantation, 62
transport, 13, 57, 82, 132, 134, 157, 163, 173
trauma, 6
treatment, x, 17, 22, 31, 47, 49, 50, 59, 60, 62, 71, 75, 82, 83, 90, 91, 99, 110, 114, 115, 118, 120, 121, 123, 129, 131, 133, 154, 163, 181
trial, 21, 47, 75, 82
tricarboxylic acid, 134
triggers, 42, 45, 101
triglycerides, 134, 137, 147, 151
TSH, 37
tuberculosis, 147, 165
tumor(s), 8, 25, 69, 112, 141, 150, 174, 175, 181
tumor necrosis factor, 112, 175, 181
tumor progression, 150
tumorigenesis, 117
turnover, ix, 62, 102, 106, 107, 109, 138, 180
type 1 collagen, 118
type 2 diabetes, 44, 46, 141, 171, 181
tyrosine, 26, 132, 141, 142